Studies in Logic
Volume 90

Model Theory for Beginners
15 Lectures

Studies in Logic Series Editor
Dov Gabbay   dov.gabbay@kcl.ac.uk

Model Theory for Beginners
15 Lectures

Roman Kossak

ISBN 978-1-84890-361-6

College Publications
Scientific Director: Dov Gabbay
Managing Director: Jane Spurr

http://www.collegepublications.co.uk

Contents

Preface

Model theory is a relatively young and rapidly developing mathematical discipline. While some of its important results date back to the 1920's, it became a well defined field of study only in the 1950's. A comprehensive history still waits to be written, but for a very informative outline see *A short history of model theory* by Wilfrid Hodges in [**BW18**].

For many years, I have taught model theory at the Graduate Center of the City University of New York. One thing I quickly learned was that even for graduate students who are well prepared in other areas of mathematics, but who had no prior exposure to mathematical logic, introduction to model theory is not easy. The problem seems to be not that the material is technically difficult, but rather that mathematical logic offers a different perspective on topics that may otherwise be familiar. We talk about the usual mathematical objects and structures, but now a close attention is paid to the language in which definitions and arguments are formalized. It takes a while for students to see the difference between ordinary mathematical discourse that is rigorous but not formal and arguments involving logical formalism. Suddenly, not everything is allowed. What makes it even more bemusing, is that after careful, often pedantic, presentations of preliminary material, texts on model theory move to what is also the usual practice in other areas of mathematics: informal rigor replaces fully formal explanations, tacit assumptions are commonly made, sometimes with little notice, abuses of notation are ubiquitous. None of it matters much to the researchers, in fact, without all those special conventions and shortcuts, research articles would be almost impossible to read. However, for students it is all rather difficult and one encounters the same difficulties trying to explain applications of model theory to fellow mathematicians from other disciplines.

In these notes, I selected topics and organized the material for an introduction to model theory that is much slower than in most textbooks, keeping in mind both kinds of readers, beginning graduate students for whom the notes can serve as preparation for more advanced topics, as well as working mathematicians who would like to quickly learn the basics that are necessary to appreciate the uses of the model-theoretic methods when applied to analysis, algebra, or combinatorics.

There is an extensive literature of the subject. There is a classic monograph by Chen Chung Chang and H. Jerome Keisler [**CK90**] that covers

most important developments until the 1980's. There is a later, encyclopedic
Model Theory by Wilfrid Hodges [**Hod93**] and its abbreviated textbook version
[**Hod97**]. There are excellent and tested in practice textbooks, in particular
[**Doe96, Mar02, Poi00, Rot00**]. There are more recent texts [**Kir19, TZ12**]
covering more up to date material. There is an interesting perspective on
mathematical logic and model theory by Yuri Manin, with an appendix on
recent developments in model theory written by Boris Zilber [**Man10**]. Jouko
Väänänen's [**Vää11**] covers much of modern model theory in the framework of
game theoretic semantics.

Two books, Maria Manzano's [**Man99**] and the recent [**Kir19**] by Jonathan
Kirby, are undergraduate and master level texts in model theory targeting similar audience. This book cover much less material, concentrating instead on
three major themes: definability, classification of types, and the use of automorphisms. David Marker's article in [**GBGL08**] gives a very good overview
of the subject.

0.0.1. The content. The basic topics and concepts discussed here are:
first-order logic, Tarski's definition of truth, first-order definability, elementary theories of structures, elementary extensions, types, indiscernible elements,
minimality and o-minimality, categoricity, saturation, and stability. Most examples are familiar: simple graphs, ordered sets of numbers, additive and multiplicative structure of the integers and the reals, and the standard model of
arithmetic. The model-theoretic analysis of these structures is not always easy,
in fact it often is not, but the book presents only those results that can be
given relatively easy proofs, using only the material explained in the book in
detail.

Special attention is given to automorphisms and their use for proving results on definability and classification of types in structures. Much of the
material in the book was developed when I was thinking of what is needed to
explain the following theorem: the ordered set of natural numbers is minimal,
i.e., the only sets of natural numbers that are definable in this structure are
either finite or cofinte. A simple proof, using an elementary extension of the
natural numbers and its automorphism in given in Chapter 8.

Up to Chapter 7, the presentation follows a more or less what is standard
in model theory textbooks and it culminates with a proof of the compactness
theorem. It diverges after that. The aim was to cover some selected topics
that also can be found in many textbooks, but to cover them much earlier
and limiting the scope to what can be done with "bare hands," i.e., using,
with some rare exceptions, only the material developed in previous chapters.
In particular, I tried to emphasize the importance for applications of the fact
that every structure with an infinite domain has a proper elementary extension.
This simple consequence of the compactness theorem is a workhorse of model
theory.

The compactness theorem is a fundamental tool of model theory. A proof
that is given in Chapter 7 works only for countable languages. For the full

version, one needs a set theoretic background with arguments involving either Zorn's lemma or some discussion of ultrafilter on arbitrary sets. I tried to limit the use of set theory to a minimum and this minimum can be described as the intuitive understanding of arguments involving constructions in infinitely many steps. In other words, the proof of the compactness theorem given here proceeds by the usual mathematical induction; the full proof requires a version of the axiom of choice. A similar approach is applied to several other main results. They usually hold in greater generality, but often, even to formulate their general version, a higher level of set theory is needed. Familiarity with axiomatic set theory is not a prerequisite for a course based on these lectures.

The compactness theorem is one of the two *big* theorems in model theory discussed here. The other is a theorem of Andrzej Mostowski and Andrzej Ehrenfeucht on existence of models admitting nontrivial automorphisms. All other material serves just one purpose, to illustrate the basic concepts and to show how they are applied. Many small results are included. Most of them can be found in other texts, although, they are usually buried in remarks in exercises sections among much more advanced material. Those results are not benchmarks of model-theory, but rather mathematical anecdotes. Some small results you might mention to a colleague at lunch. An example: multiplication has no first-order definition in the additive structure of the natural numbers.

Chapters 8 and 12 are a bit more advanced. Chapter 8 finishes with a proof of the infinite Ramsey theorem for pairs that uses an elementary extension of the standard models. The main result in Chapter 12 is Abraham Robinson's model-theoretic proof of Tarski's undefinability of truth theorem. The proof is preceded by Skolem's construction of a nonstandard model of arithmetic. The material is more advanced, but still more accessible than what can be found in the literature and that is because the results are not given in their full generality. While the results mentioned here hold for models of Peano Arithmetic, here they are only formulated for models of True Arithmetic, i.e., the first-order theory of the standard model of arithmetic. The same applies to a short discussion of recursively saturated models of arithmetic in Chapter 13.

In the opinion of many model-theorists, model theory begins in earnest in the 1960's with Morley's categoricity theorem and subsequent work of Saharon Shelah that he developed into what is known today as stability theory. *Stability* is the title of the last, very short, chapter and it is included to indicate that there is much more to model theory than what is presented in these lectures and also to give another example of how techniques discussed in this book are used.

Hundreds of interesting exercises can be found in any of the textbooks I mentioned above. Exercises are also included here, but there are no exercise sections. Exercises are scattered throughout the text. The reader is invited to do them at the time when they are reading the particular sections to which they are attached.

Thank you note. The lecture notes presented here are an expanded version of the notes I prepared for a two week mini-course that I gave at Nesin Mathematics Village and at the National Technical University of Athens in September and October of 2017. I want to thank Costas Dimitracopoulos, Ali Nesin, and Stathis Zachos for their support. Alfred Dolich, Stephen Kennedy, Anthony Weaver, and my former student Jordi Navarrette have read preliminary versions of this book. I am very grateful for their comments and corrections.

I am very grateful to Costas Dimitracopoulos, Fredrik Engsrtröm and Alex Kampa for careful reading of the first printed version of the book and providing errata that I used for this corrected version.

CHAPTER 1

Structures

Model theory can be described as a general theory of mathematical structures, so, we will begin with a short discussion of what makes a structure
mathematical. In wikipedia one can read: *In mathematics, a structure is a
set endowed with some additional features on the set (e.g., operation, relation,
metric, topology).* To simplify matters, we will first introduce in Definition
1.14 relational structure. This definition is particularly simple. All mathematical structures discussed here can be represented as relational, although
for algebraic structures such as groups or vector spaces, this requires some ex-
tra effort. In later lectures, we will extend the definition to be able to treat
algebraic structures in more natural ways.

Before giving a general definition of a relational structure, let us consider
some examples that will motivate that definition.

1.1. Digraphs

A *directed graph*, or *digraph* for short, consists of a set of vertices and a
set of directed edges. For each pair of vertices, either there is an edge pointing
from one vertex to the other, or there isn't and there can be edges pointing in
both directions.

For example, let G be the digraph with vertices a, b, c, and d and edges
$a \mapsto b$, $a \mapsto c$, and $a \leftrightarrow d$.

The definition of digraphs in terms of vertices and edges is perfectly fine.
It tells us what digraphs are and at the same time introduces the terminology
that is used to talk about them. Still, the following definition introduces the
same class of structures in different terms.

DEFINITION 1.1. A *digraph* is an ordered pair (V, E), where V is a set and
E is a set of ordered pairs of elements of V.

The digraph G in our example above is the set of vertices $V = \{a, b, c, d\}$,
with the set of edges $E = \{(a, b), (a, c), (a, d), (d, a)\}$, or that G is the ordered
pair (V, E). This way, G is represented as a purely set-theoretic object.

What happens here is simple. When talking abut digraphs, we still call
their elements vertices and ordered pairs of vertices edges, but we do not need
to explain what they are; they are just a set V and a set of E ordered pairs
of elements of V. This gives us a very general definition: *Any* set can serve as
the set of vertices of a digraph and *any* set of ordered pairs of elements of that

set can be a set of edges. The definition in such a generality puts digraphs in the same category as all other structures that model theory deals with; hence the general techniques apply to digraphs in particular.

Definition 1.1 allows the set E to be empty. Such digraphs are just sets of unconnected vertices. At the other extreme are the *complete digraphs* whose edges are all ordered pairs (w, v), with v and w in V.

Digraphs can have edges connecting a point to itself. Such edges are represented by pairs (v, v) and are called *loops*.

A very broad goal of model theory is to classify structures, or at least to classify some structures according to some criteria, at least partially. For this task, it is very important to realize how many structures of a given kind there are.

How many different digraphs on the set of n vertices are there? Let us count them for $n = 3$, assuming that there are no loops. For each digraph, we need to choose a set of edges. Let $V = \{a, b, c\}$. The set of all possible edges is

$$\{(a, b), (b, a), (a, c), (c, a), (b, c), (c, a)\}.$$

This set has $2^6 = 64$ subsets, so this is the number of all digraphs with the set of vertices V. It is a lot. If E_1 and E_2 are different sets of edges, then (V, E_1) and (V, E_2) are different, but they may not differ much. For example, $(V, \{(a, b)\})$ and $(V, \{(a, c)\})$ are different, but they look the same. The idea of "looking the same" is captured by the notion of *isomorphism*.

DEFINITION 1.2. Graphs $G_1 = (V_1, E_1)$ and $G_2 = (V_2, E_2)$ are *isomorphic* if there is a one-to-one and onto function $f : V_1 \longrightarrow V_2$ such that for all v, w in V_1, (v, w) is in E_1 if and only if $(f(v), f(w))$ is in E_2.

In our example above $V_1 = V_2 = V$ and the isomorphism is defined by $f(a) = a$, $f(b) = c$, and $f(c) = b$.

EXERCISE 1.3. Let $V = \{a, b, c\}$, $E_1 = \{(a, b), (b, c)\}$, $E_2 = \{(b, c), (c, a)\}$, and $E_3 = \{(a, b), (c, b)\}$. Show that (V, E_1) and (V, E_2) and isomorphic, but (V, E_1) and (V, E_3) are not.

EXERCISE 1.4. Find the formula for the number of different digraphs with n vertices and compute it for small values of n.

The numbers of different digraphs even for relatively small sets of vertices are staggering, but we are more interested in the numbers of digraphs that are really different, i.e., the number of nonisomorphic ones. Those numbers are much harder to find. There are no known effective algorithms to decide whether two given digraphs on sets of vertices of the same finite size are isomorphic. The search for such algorithms is a whole mathematical discipline.

1.2. Ordered Sets

DEFINITION 1.5. A *linear ordering* is a relation $<$ on a set A which satisfies the following three conditions for all a, b, and c in A,

(1) $a < b$ or $b < a$ or $a = b$;

(2) $a \not< a$;

(3) if $a < b$ and $b < c$, then $a < c$.

Linear orderings are also called *total* orderings.

Conditions (2) and (3) imply that for all a and b only one of $a < b$, $b < a$, $a = b$ holds. Instead of "linear ordering" and "linearly ordered sets," we will just say "ordering" and "ordered set."

The order relation $<$ can be represented as a set of ordered pairs (a, b) such that $a < b$; hence ordered sets are a special kind of digraphs.

A *permutation* of a set X is a function $f : X \longrightarrow X$ that is one-to-one and onto.

Let $A = \{a_1, \ldots, a_n\}$. For a permutation f of $\{1, \ldots, n\}$, let the ordering $<_f$ of A be defined by

$$a_i <_f a_j \text{ if and only if } f(i) < f(j).$$

This gives us all possible $n!$ orderings of A. There are many orderings, but they all look the same. They are all isomorphic to the ordered set $\{1, \ldots, n\}$ with its usual ordering. While there are many orderings of a set of size n, up to isomorphism there is just one. The situation changes when we consider orderings of infinite sets.

Note on notation: We will use $\mathbb{N}$ for the set of natural numbers $\{0, 1, 2, \ldots\}$; $\mathbb{Z}$ for the set of integers, $\mathbb{Q}$ for the set of rational numbers, $\mathbb{R}$ for the set of real numbers, and $\mathbb{C}$ for the set of complex numbers.

The three ordered sets $(\mathbb{N}, <)$, $(\mathbb{Z}, <)$, and $(\mathbb{Q}, <)$, with the usual ordering $<$, are not isomorphic to one another. In $(\mathbb{N}, <)$ there is a least element and there is no such element in the other two sets. $(\mathbb{N}, <)$ and $(\mathbb{Z}, <)$ are *discrete* orderings, which means that for every number n there are no numbers between n and its *successor* $n + 1$. The ordering of $(\mathbb{Q}, <)$ is *dense*, for all distinct rational numbers p and q, there is a rational number between them. It doesn't seem that one could say much more about these ordered structures, but we will still analyze them in great detail later.

EXERCISE 1.6. Ordered sets are directed graphs. Use the discussion in the previous paragraph to show that the structures $(\mathbb{N}, <)$, $(\mathbb{Z}, <)$ and $(\mathbb{Q}, <)$ are pairwise nonisomorphic according to Definition 1.2.

1.3. Groups and Fields

Much of model theory can be seen as a generalization of abstract algebra to the case of arbitrary mathematical structures. Algebra deals with analyzing and classifying algebraic structures, often by means of their substructures and extensions of various kinds. We will use groups and fields as examples.

DEFINITION 1.7. A *group* is a structure $(G, \circ, e)$, where G is a set, $\circ$ is a binary operation on G and e is an element of G, called the *identity element* of the operation $\circ$, satisfying the following conditions:

(1) For all a, b, and c, $a \circ (b \circ c) = (a \circ b) \circ c$.
(2) For all a, $a \circ e = a$, and $e \circ a = a$.
(3) For each a there is a b such that $a \circ b = e$.

The smallest group consists of one element e and $e \circ e = e$. It is the *trivial group*.

In Definition 1.7, e and $\circ$ are generic symbols for an identity element and a group operation. In specific examples more familiar symbols are used.

The group $\mathbb{Z}_2$ with two elements 0, and 1, has its binary operation $+$ defined by $0 + 0 = 0$, $0 + 1 = 1 + 0 = 1$, and $1 + 1 = 0$. The identity element is 0.

EXERCISE 1.8. Verify that the trivial group and $\mathbb{Z}_2$ are groups according to Definition 1.7.

$(\mathbb{Z}, +, 0)$ is a group, but $(\mathbb{Z}, \cdot, 1)$ is not, it does not satisfy condition (3) in Definition 1.7.

EXERCISE 1.9. Show that $(\mathbb{Q}, +, 0)$ and $(\mathbb{Q}^+, \cdot, 1)$ are groups, where $\mathbb{Q}^+$ is the set of the positive rational numbers.

DEFINITION 1.10. A *field* is a structure $\mathfrak{F} = (F, +, \cdot, 0_{\mathfrak{F}}, 1_{\mathfrak{F}})$, where $+$ and $\cdot$ are binary operations and $0_{\mathfrak{F}}$ and $1_{\mathfrak{F}}$ are distinct elements of F, satisfying the following conditions:

(1) $0_F \neq 1_{\mathfrak{F}}$.
(2) $(F, +, 0_{\mathfrak{F}})$ is a group such that for all a, b in F, $a + b = b + a$.
(3) $(F^*, \cdot, 1_{\mathfrak{F}})$ is a group such that for all a, b in F, $a \cdot b = b \cdot a$, where F^* is the set of all nonzero elements of F.
(4) For all a, b, and c, $a \cdot (b + c) = a \cdot b + a \cdot c$.

$(\mathbb{Q}, +, \cdot, 0, 1)$ is a field, but because $(\mathbb{Z}^*, \cdot, 1)$ is not a group, $(\mathbb{Z}, +, \cdot, 0, 1)$ is not a field.

The smallest field, usually denoted $\mathbb{F}_2$, is $(\{0, 1\}, +, \cdot, 0, 1)$, where $(\{0, 1\}, +, 0)$ is $\mathbb{Z}_2$ and $(\{0, 1\}^*, \cdot, 1)$ is the trivial one element group.

EXERCISE 1.11. Verify that $\mathbb{F}_2$ is a field.

1.4. Relational Structures

For a set A, A^1 is A, A^2 is the set of all ordered pairs of elements of A, A^3 is the set of all ordered triples. This is generalized as follows. For a natural number $n > 0$, an *n-tuple* is a sequence of length n.

DEFINITION 1.12. Let n be a positive natural number. The *n-ary Cartesian power* of a set A, denoted A^n, is the set of all n-tuples of elements of A. By $A^{<\omega}$ we denote $\bigcup_{n>0} A^n$.

The next two definitions are crucial. In the short discussion so far, the material was introduced as is routinely done in mathematics. The concepts are precise, but their descriptions are often based on intuitive understanding of basics such as sets, relations, and operations (functions). In model theory, the aim often is to identify certain properties of *all* structures and to classify them *all* according to some criteria involving those properties. Together with basic intuitions, we also need more precision. What is a relation? What is a structure? These questions are answered in the following two definitions. They are highly abstract. Everything is reduced to only one primitive notion of *set*.

In Definition 1.12, we also refer to n-tuples. This could be a primitive notion as well, but n-tuples can be represented as sets. First, we define the unordered pair of elements a and b to be the set $\{a, b\}$. In particular, $\{a, b\} = \{b, a\}$ and $\{a, a\} = \{a\}$. Then, we define the ordered pair (a, b) to be the Kuratowski's pair $\{a, \{a, b\}\}$. Using basic axioms of set theory one can show that for all a, b, c, and d, $\{a, \{a, b\}\} = \{c, \{c, d\}\}$ if and only if $a = c$ and $b = d$.

The triple (a, b, c) is identified with the pair $((a, b), c)$ and , by induction, the $n + 1$-tuple $(a_1, \ldots, a_n, a_{n+1})$ with $((a_1, \ldots, a_n), a_{n+1})$. This is all we need to turn all objects that we will define below into sets.

What is a set then? That question does not have a straightforward answer. For most uses in mathematics, our intuition serves us well, but there are complications when sets of large sizes and issues such as the axiom of choice or the continuum hypothesis emerge. For more advanced developments, we need to refer to axiom systems. They do not tell us what sets are, but describe the properties of what is called the universe of sets, or, in fact, quite diverse universes of all sets. Set theory impacts model-theoretic developments to a large extent, but those are exactly the developments that will not be discussed in these lectures.

DEFINITION 1.13. For a positive natural number n, an *n-ary relation* on a set A is a subset of A^n. If R is an n-ary relation, then n is called the *arity* of R. A *relation* is an n-ary relation for some n.

DEFINITION 1.14. A *relational structure* is a non-empty set A together with a set of relations on it. The set A is called the *domain* of the structure.

Digraphs are relational structures, groups and fields are not; they are defined in terms of operations (functions). However, every such structure can be represented as a relational structure as follows.

Note on notation: If $f : A \longrightarrow B$ is a function, then for $\bar{a} = (a_1, \ldots, a_n)$ in $A^{<\omega}$, $f(\bar{a})$ denotes $(f(a_1), \ldots, f(a_n))$.

For any set A and a function $f : A^n \longrightarrow A$ for $n > 0$, let the $(n + 1)$-ary relation R_f be the *graph* of f, defined by

$$(\bar{a}, b) \in R_f \text{ if and only if } f(\bar{a}) = b,$$

for all $\bar{a}$ in A^n, and all $b \in A$.

Because all functions can be represented as relations, all mathematical structures can be represented as relational structures. This comes at a cost, as some natural arguments involving functions became more cumbersome in the relational language. For that reason, later we will have to expand Definition 1.14 to let the functions in.

EXAMPLE 1.15. The ternary relation that represents addition in $\mathbb{Z}_2$ is the set

$$\mathcal{A}_{\mathbb{Z}_2} = \{(0,0,0),(0,1,1),(1,0,1),(1,1,0)\}.$$

Constants play important role and—as we have seen—they are considered parts of structures. They can be represented as one element sets and sets are also relations, since we identify A^1 with A. We call such relations *unary*. With this convention in mind, the two element group $\mathbb{Z}_2$ can be represented as the relational structure $(\{0,1\}, \mathcal{A}_{\mathbb{Z}_2}, \{0\})$.

Note on notation: For names of structures, we will use Math Fraktur font, typically $\mathfrak{M}$ and $\mathfrak{N}$ and for their domains we will use the same characters in regular math font. So, M and N are the domains of structures $\mathfrak{M}$ and $\mathfrak{N}$, respectively.

If $\mathfrak{M}$ is a structure and $\mathcal{R}$ is a set of relations on M that are not relations of the structure $\mathfrak{M}$, the structure obtained by including those relations is an *expansion* of $\mathfrak{M}$, and we denote it by $(\mathfrak{M}, \mathcal{R})$. We call $\mathfrak{M}$ a *reduct* of $(\mathfrak{M}, \mathcal{R})$. Every structure is an expansion of its domain and its domain is its reduct.

1.5. Small Technical Difficulties.

All classical structures of mathematics can be represented as relational structures as described above. Let us take a look at two types of structures for which it takes some effort.

1.5.1. Vector Spaces. A vector space over a field $\mathfrak{F}$ is a group $(V, +)$, in which the group operation is commutative, i.e., for all v and w in V, $v + w = w + v$, equipped with an operation of multiplying the elements of V (vectors) by the field elements (scalars). The axioms for multiplication are: for all field elements a, b, and c and all vectors v and w:

(1) $1_{\mathfrak{F}} v = v$;
(2) $a(bv) = (ab)v$;
(3) $a(v + w) = av + aw$;
(4) $(a + b)v = av + bv$.

There are two additions and two multiplications involved. For example, in $a(bv)$ there are two scalar multiplications. First v is multiplied by b and then the resulting vector is multiplied by a. In $(ab)v$, first a is multiplied by b in the field and then v is multiplied by the resulting scalar.

There are alternative ways to represent vector spaces in a model-theoretic fashion. One is to consider the structure whose domain is the disjoint union of V and the domain F and to consider the vector space operations as operations on the space of ordered pairs of elements V (addition) and of ordered pairs (a, v) (scalar multiplication), where a is in F and v is in V. Moreover, the addition and multiplication of $\mathfrak{F}$ are also part of the structure. This is straightforward, but it makes the structure much more complex than it needs to be for the model-theoretic analysis. The reason is that in this representation, the field is a part of the structure. Fields can be very complex while vector spaces are much simpler to analyze. To avoid this problem, we need another representation.

Instead of incorporating $\mathfrak{F}$ into the structure of the vector space, we can consider the group $(V, +)$ expanded either by adding the functions $f_a : V \mapsto V$, representing multiplication by a, one for each a in F, or, to get a relational structure, by binary relations R_a on V defined by $(v, w) \in R_a$ if and only if $av = w$.

1.5.2. Graphs. A *simple graph*, or just *graph*, is a set of vertices, with a set of undirected edges. Undirected edges are unordered pairs of vertices $\{w, v\}$. Each such graph G can be represented as a digraph by identifying each undirected edge $\{w, v\}$ in G with the pair of directed edges (v, w) and (w, v). This is a bit awkward, but not a problem. This is a common feature of *symmetric* relations, i.e., binary relations R such that for all a, b in the domain, (a, b) is in R if and only if (b, a) is.

Graphs with multiple edges between the same vertices are common in applications. Our definitions do not cover such graphs directly, but there are several ways to represent them. Here is one.

Let G be a digraph with three vertices $V = \{a, b, c\}$ and three edges $E = \{(a, b), (b, c), (b, c)\}$. The problem here is that in set theory $\{(a, b), (b, c), (b, c)\} = \{(a, b), (b, c)\}$. So instead, we let E be the set of edges $\{e_1, e_2, e_3\}$. We can now represent G as the structure with domain $V \cup E$, two unary relations (sets) V and E and one ternary relation $\{(e_1, a, b), (e_2, b, c), (e_3, b, c)\}$.

CHAPTER 2

Language

Mathematical structures are studied by a variety of means. In these lectures we will look at structures from the vantage point of mathematical logic, and we will pay close attention to the language in which structures can be described and characterized.

In mathematics, we use precise definitions and our arguments have rigorous logical structure. In the previous chapter we used symbols and mathematical formulas, but the language was still the ordinary language of mathematics that is to a large extent informal. The formal language of first-order logic that we will define in this chapter is a mathematical construct.

2.1. Symbols

Here is a list of logical symbols of first-order logic:

(1) Logical connectives: $\wedge$ (for "and") and $\neg$ (for "not").
(2) Quantifiers: existential $\exists$ (for "there is").
(3) Variables: x_i, where i is a natural number.
(4) The equality relation symbol $=$.
(5) Auxiliary symbols: left and right parentheses and comma.

The list above is not complete, we will also use symbols $\vee$ for "or", $\implies$ for "if ... then", $\iff$ for "if and only if", and $\forall$ for universal quantification, but since all these connectives and the universal quantifier can be defined in terms of the symbols listed above, we will keep the official list short to simplify some definitions and proofs.

To express properties of particular structures, we need to add relation symbols, one for each relation of the given structure or collection of structures. We will call this selection of symbols the *language* of the structure or a class of structures. The language of a structure is also called its *signature*.

As mentioned in the previous chapter, all initial results will be formulated in terms of relational structures. Presence of functions makes statements of some of those results more natural, especially in algebra. To formalize such statements we use function symbols. There is no problem with including function symbols and this is routinely done, but at the cost of some notational complications.

The only relation of a graph is the edge relation, represented by a set of ordered pairs of the vertices of the graph. Thus, to express properties of graphs in first-order logic, we only need one relation symbol. Let it be $\mathcal{E}$.

For a graph $G = (V, E)$, the most basic information is whether for a pair of vertices v and w there is an edge between them or not. In the mathematical language we write $(v, w) \in E$ if there is an edge between the vertices and $(v, w) \notin E$ if there is not.[1] To express that (v, w) is in E in the formal language of first-order logic, we will do something that initially may seem to be excessively pedantic. We will do it by saying that the *formula* $\mathcal{E}(x_1, x_2)$ is *satisfied* in G, when x_1 and x_2 are evaluated as v and w, respectively. Let us repeat the same definition using symbolic notation.

DEFINITION 2.1. For a graph $G = (V, E)$ and v and w in V,

$$G \models \mathcal{E}(x_1, x_2)[v, w] \text{ if and only if } (v, w) \in E. \tag{*}$$

What is written on the right side of the equivalence (*) above is clear. The left side needs an explanation. It says that that $\mathcal{E}(x_1, x_2)$ is *satisfied* in G by the evaluation of variables that assigns v to x_1 and w to x_2. The statement on the left is about a ternary relation involving G, the pair of its vertices (v, w) and the formula $\mathcal{E}(x_1, x_2)$; the statement on the right is a condition that determines whether that relation holds or not. The graph and the pair of its vertices are ordinary mathematical objects; $\mathcal{E}(x_1, x_2)$ is also a mathematical object, but of a different kind. It is a finite string of symbols of first-order logic.

In some structures, particular elements play a special role. We call those elements *constants*. Constants can be represented in relational structures as one-element sets, but to refer to constants in a formal way, it will be more convenient to use *constant symbols*. Typically they would be symbols endowed with meaning such as 0 and 1, but any symbol can be used.

Let us repeat, the *language* of a structure is the set of non-logical symbols for the relations of the structure. For example, the language that we chose for graphs is just the one element set $\{\mathcal{E}\}$.

2.2. Formulas

We will first define the formulas for the language of graphs and then we will generalize to arbitrary structures.

DEFINITION 2.2. *Atomic formulas* of the language of graphs are expressions of the form $x_i = x_j$ and $\mathcal{E}(x_i, x_j)$, for all i and j in $\mathbb{N}$.

All other first-order formulas of the language of graphs are built from the atomic ones according to the rules below.

DEFINITION 2.3. The set of first-order formulas of the language of graphs is the smallest set $\mathsf{Form}_{\mathcal{E}}$ such that,

---

[1] In set theory, $\in$ is the membership relation symbol; $a \in X$ means that a is in the set X and $a \notin X$ that it is not.

(1) all atomic formulas are in $\mathsf{Form}_\mathcal{E}$;
(2) if φ and ψ are in $\mathsf{Form}_\mathcal{E}$, then so is $(\varphi) \wedge (\psi)$;
(3) if φ is in $\mathsf{Form}_\mathcal{E}$, then so is $\neg(\varphi)$;
(4) for each variable x_i, if φ is in $\mathsf{Form}_\mathcal{E}$, then so is $\exists x_i(\varphi)$.

The conditions (2), (3), and (4) are the only rules for generating formulas. The requirement that $\mathsf{Form}_\mathcal{E}$ is the smallest set satisfying the conditions of Definition 2.3 means that only formulas generated by applications of those rules are in $\mathsf{Form}_\mathcal{E}$.

EXAMPLE 2.4. Here are three examples of well-formed first-order formulas.
(1) $\exists x_1(\mathcal{E}(x_1, x_2))$ says that there is a vertex that forms an edge with x_2.
(2) $\neg(\exists x_1(\mathcal{E}(x_1, x_1)))$ says that there are no loops.
(3) $\exists x_2(\exists x_3((\neg(x_1 = x_2)) \wedge ((\neg(x_1 = x_3)) \wedge ((\neg(x_2 = x_3)) \wedge (\mathcal{E}(x_1, x_2)) \wedge (\mathcal{E}(x_2, x_3))))))$ says there is a path made of two edges, from x_1 to another vertex.

EXERCISE 2.5. Verify that all examples above are properly formed formulas according to Definition 2.3.

According to Definition 2.3 a new pair of parentheses is needed every time a rule for generating formulas is applied. That makes longer formulas almost unreadable, but in practice, if there is no danger of confusion, we will write formal expressions without them. For example, (3) above is usually written this way:

$$\exists x_2 \exists x_3 [\neg(x_1 = x_2) \wedge \neg(x_1 = x_3) \wedge \neg(x_2 = x_3) \wedge \mathcal{E}(x_1, x_2) \wedge \mathcal{E}(x_2, x_3)].$$

In example (1), the variable x_1 is *quantified* by $\exists$. The other variable x_2 is not quantified. We call unquantified variables *free*. In (3), x_2 and x_3 are quantified, x_1 is free. In (2) there are no free variables, such formulas are called *sentences*.

Definition 2.3 says nothing about the meaning of formulas. There is nothing that prevents us from forming formulas that do not make much sense, for example $\exists x_{1000}\mathcal{E}(x_1, x_2)$, or $(\varphi) \wedge ((\varphi) \wedge (\varphi)))$. That feature is not a deficiency. It is important that first-order formulas can be generated by a mechanical process without any regard to their interpretations.

2.3. Satisfaction

In this section we will define what it means for a sentence in the language of a structure to be true in the structure. The definition is crucial for the development of model theory and it has an interesting history. Wilfird Hodges writes about it in the Stanford Encyclopedia of Philosophy [**Hod18**]. Here is a quotation:

> In 1933 the Polish logician Alfred Tarski published a paper
> in which he discussed the criteria that a definition of "true

sentence" should meet and he gave examples of several such definitions for particular formal languages. In 1956 Tarski and his colleague Robert Vaught published a revision of one of the 1933 truth definitions, to serve as a truth definition for model-theoretic languages.

In Tarski's definition, rather than defining truth values (true or false) of a sentence directly, we will first define satisfaction of formulas by evaluations of their free variables by elements from the domain of a structure. It is all rather straightforward, but requires some care with notation. For that reason, we will describe it first in detail using examples with particularly simple languages and then we will just describe how to modify the examples to get a general definition. It will also help that we will now define satisfaction only for relational structures. The definition can be generalized to structures involving functions with no problem, but at a cost of adding another layer of notational complexity.

Satisfaction for atomic formulas of the language of graphs has already been given in Definition 2.1. Now we will repeat it a bit more precisely and then we will extend the definition to all formulas of the language of graphs.

We have an infinite supply of variables x_i, one for each natural number i. Each formula has only finitely many free variables, but in order to uniformly define satisfaction for all of them we need evaluations that assign an element of the domain of a given structure to each variable.

For a graph (V, E), an *evaluation* is a function $\alpha : \{x_i : i \in \mathbb{N}\} \longrightarrow V$. For each evaluation α and for any vertex v, $\alpha_{(x_n/v)}$ is the evaluation defined by

$$\alpha_{(x_n/v)}(x_i) = \begin{cases} \alpha(x_i), & \text{if } i \neq n, \\ v, & \text{if } i = n. \end{cases}$$

We will now define what it means that φ is satisfied in G by an evaluation α, written $G \models \varphi[\alpha]$.

DEFINITION 2.6. Let $G = (V, E)$ be a graph, let $\alpha : \{x_i : i \in \mathbb{N}\} \longrightarrow V$ be an evaluation, and let ψ and θ be formulas of the language of graphs. Then, for all i and j

(1) $G \models (x_i = x_j)[\alpha]$ if and only if $\alpha(x_i) = \alpha(x_j)$, and $G \models \mathcal{E}(x_i, x_j)[\alpha]$ if and only if $(\alpha(x_i), \alpha(x_j)) \in E$.

(2) $G \models ((\psi) \wedge (\theta))[\alpha]$ if and only if $G \models \psi[\alpha]$ and $G \models \theta[\alpha]$.

(3) $G \models \neg(\psi)[\alpha]$ if and only if it is not the case that $G \models \psi[\alpha]$.

(4) $G \models \exists x_i(\psi)[\alpha]$ if and only if there is v in V such that $G \models \psi[\alpha_{(x_i/v)}]$.

The definition of $G \models \varphi[\alpha]$ is by *induction on the complexity* of φ. Part (1) covers the case of atomic φ. If φ is not atomic, then it is composed of its subformulas according to Definition 2.3. For example, if φ is of the form $\psi \wedge \theta$, then in the definition we inductively assume that $G \models \psi[\alpha]$ and $G \models \theta[\alpha]$ have been defined. This allows us to define $G \models \varphi[\alpha]$ as in (2). To find out whether or not $G \models \varphi[\alpha]$, we can keep decomposing φ into its subformulas until we get down to the level of atomic statements. Then, we can check directly which of the atomic statements are true and which are false. After that we can recover the status of φ by reconstructing the formula and keeping track of the satisfaction relation for its subformulas.

When we say that we can check whether an atomic formula, say $\mathcal{E}(x_1, x_2)$, is satisfied by an evaluation that assigns vertices v and w to x_1 and x_2, respectively, by checking if the graph has an edge from v to w, we do not mean that there actually is an effective procedure to find out if this is the case. We only mean that we know the condition under which the formula is satisfied by an evaluation. This feature of the definition of satisfaction is often explained by the example: the statement "it is snowing" is true if and only if it is snowing (regardless of whether we know if it is snowing now).

Definitions 2.3 and 2.6 were given for the language of graphs. To define formulas and the satisfaction relation for all relational structures, only small modifications are needed. The definitions are the same except that part (1) in Definition 2.6 applies to atomic formulas involving all relation symbols of the language and, in addition, if the language includes constant symbols, then the set of atomic formulas is extended to include formulas of the form $x_i = c$, $c = x_i$, and $c = d$, for all variables x_i and all constant symbols c and d. If c is a constant symbol, then to (1) we add the condition

$$\mathfrak{M} \models (x_i = c)[\alpha] \text{ if and only if } \alpha(x_i) = c^{\mathfrak{M}},$$

where $c^{\mathfrak{M}}$ is the interpretation of c in $\mathfrak{M}$

EXAMPLE 2.7. The language of groups consists of one ternary relation symbol $\mathcal{O}$ and one constant symbol e. Let $\mathfrak{G} = (G, \circ, e^{\mathfrak{G}})$ be a group. Notice that now we use $e^{\mathfrak{G}}$ for the neutral element of G to distinguish it from the formal symbol e. If $\alpha : \{x_i : i \in \mathbb{N}\} \longrightarrow G$ is an evaluation, then (1) in Definition 2.6 is replaced by the following two conditions. For all natural numbers i, j, and k,

$$\mathfrak{G} \models \mathcal{O}(x_i, x_j, x_k)[\alpha] \text{ if and only if } \alpha(x_i) \circ \alpha(x_j) = \alpha(x_k),$$

and

$$\mathfrak{G} \models (x_i = e)[\alpha] \text{ if and only if } \alpha(x_i) = e^{\mathfrak{G}}.$$

EXERCISE 2.8. Choose the language of fields to be two ternary relation symbols $\mathcal{A}$ and $\mathcal{M}$, for addition and multiplication respectively, and two constant symbols 0 and 1, whose interpretations in the given field $\mathfrak{F}$ are $0^{\mathfrak{F}}$ and $1^{\mathfrak{F}}$. Write part (1) of Definition 2.6 for fields.

EXERCISE 2.9. Write the four axioms for vector spaces from Section 1.5.1 in the relational language with relation symbols R_a for all field elements a.

2.3.1. Higher-Order Logics. According to Definition 2.6, the variables can only be interpreted as individual elements. In first-order logic we cannot interpret them as sets, functions, pairs of elements, etc. It does not mean that mathematical logic forbids such interpretations. On the contrary, there are formal systems which employ second-order variables that are interpreted as subsets of domains and their Cartesian powers. There are systems allowing third-order variables, interpreted as sets of sets. There are higher-order systems. None of those systems will be discussed in this book. For an account of second-order logic and its applications in the foundational study of classical mathematics see [**Sim09**].

2.4. Notational Conventions and Abbreviations

Our list of logical symbols of first-order logic is short. This will allow us to present basic definitions and arguments concisely. While very few symbols suffice to express everything one wants to in the first-order way, the cost is that very quickly formulas become unreadable. For that reason, let us expand the list of symbols that we will freely use. Notice that in the definition below some parentheses that are required by Definition 2.3 are omitted. We do it when it does not create ambiguity. It is standard practice.

DEFINITION 2.10. Let φ and ψ be formulas of some first-order language.
(1) Disjunction $\varphi \vee \psi$ is defined as $\neg(\neg\varphi \wedge \neg\psi)$.
(2) Implication $\varphi \implies \psi$ is defined as $\neg\varphi \vee \psi$.
(3) Equivalence $\varphi \iff \psi$ is defined as $(\varphi \implies \psi) \wedge (\psi \implies \varphi)$.
(4) Universal quantification $\forall x_i \varphi$ is defined as $\neg\exists x_i \neg\varphi$.

In later chapters, we will also use the quantifier $\exists! x$, there is exactly one x, defined by $\exists! x \varphi(x)$ if and only if $\exists x[\varphi(x) \wedge \forall y(\varphi(y) \implies y = x)]$.

In Definition 2.6 we used evaluations that assign elements of the domain of a structure to *all* variables. That was necessary, because satisfaction is defined for all formulas of the language of the structure. In practice, we usually deal with specific formulas with a fixed number of free variables. If $x_1, \ldots, x_n$ are the free variables of a formula φ and α is an evaluation such that for $i = 1, \ldots, n$, $\alpha(x_i) = a_i$, where $a_1, \ldots, a_n$ are in M, instead of

$$\mathfrak{M} \models \varphi[\alpha],$$

we will write

$$\mathfrak{M} \models \varphi(a_1, \ldots, a_n).$$

For any relation symbol $\mathcal{R}$ in the language of $\mathfrak{M}$, by $\mathcal{R}^{\mathfrak{M}}$ we denote its interpretation in $\mathfrak{M}$ and for each constant symbol c, $c^{\mathfrak{M}}$ is the element of the domain of $\mathfrak{M}$ that is named by c. With this convention, part (1) of Definition 2.6 can be written as follows. If $\mathcal{R}$ is n-ary, then, for all $a_1, \ldots, a_n$ in M

$$\mathfrak{M} \models \mathcal{R}(a_1, \ldots, a_n) \text{ if and only if } (a_1, \ldots, a_n) \in \mathcal{R}^{\mathfrak{M}}$$

and for each constant symbol c and each a in M,

$$\mathfrak{M} \models (c = a) \text{ if and only if } c^{\mathfrak{M}} = a.$$

Initially, we need to pay close attention to the separation between syntax and semantics. Formal symbols belong to syntax and they are not the same as informal names of mathematical objects they stand for. Later, we will abuse this principle and use the same characters for the formal symbols and their interpretations in structures.

EXERCISE 2.11. Recall that a formula is a sentence if it has no free variables. Use Definition 2.6 to show that if φ is a sentence and $\mathfrak{M} \models \varphi[\alpha]$ for some evaluation α, then $\mathfrak{M} \models \varphi[\alpha]$ for all evaluations α.

Because satisfaction of sentences does not depend on evaluations of variables, if φ is a sentence and it is satisfied in $\mathfrak{M}$ by some evaluation, we will just write $\mathfrak{M} \models \varphi$, and we will say that φ is *true* in $\mathfrak{M}$.

Once the syntax and semantics of first-order logic are defined, we no longer have to follow the rules concerning notation strictly. We chose x_i for variables, but that choice was arbitrary. Any other infinite set of symbols would do, and we will freely use other symbols, usually x, y, and z. Also, we will use parentheses only when ambiguity could arise. While according to Definition 2.3 the conjunction of $\mathcal{R}(x, y)$ and $\mathcal{R}(y, z)$ is $(\mathcal{R}(x, y)) \wedge (\mathcal{R}(y, z))$, instead we will just write $\mathcal{R}(x, y) \wedge \mathcal{R}(y, z)$ and instead of $\exists x(\mathcal{R}(x, y))$ we will write $\exists x \mathcal{R}(x, y)$.

Some formulas that we will consider will have long blocks of either existential or universal quantifiers. In such cases, instead of, for example, $\exists x_1 \exists x_2 \cdots \exists x_n$, we will write $\exists x_1, x_2, \ldots, x_n$, or simply $\exists \bar{x}$, where either we will specify that $\bar{x} = x_1, \ldots, x_n$, or it will be clear from the context.

CHAPTER 3

Definability

One can say that to know a structure is to know its definable sets.

DEFINITION 3.1. Let $\mathfrak{M}$ be a structure and let X be a subset of M^n. We say that X is *definable* in $\mathfrak{M}$ if there is a formula $\varphi(x_1, \ldots, x_n)$ of the language of $\mathfrak{M}$, with free variables as displayed, such that

$$X = \{(a_1, \ldots, a_n) \in M^n : \mathfrak{M} \models \varphi(a_1, \ldots, a_n)\}.$$

An element a of M is *definable* in $\mathfrak{M}$ if the set $\{a\}$ is definable in $\mathfrak{M}$.

Before we discuss definability in general, let us examine a few specific cases.

3.1. Defining Integers

Let $(A, <)$ be an ordered set. An element b in A is an *immediate successor* of a if $a < b$ and there are no elements between a and b. The ordering $<$ of a set A is *discrete* if every element of A has an immediate successor, except for the largest element, if it exists. This property can be expressed by a first-order sentence. We will do it in a moment, but we need a comment on notation.

Because ordering is a binary relation, for the language of $(A, <)$ we need a binary relation symbol. Let it be L. Then, for a and b in A,

$$(A, <) \models L(a, b) \text{ if and only if } a < b.$$

It was important in the introduction to make a strict distinction between the informal mathematical symbols and their formal counterparts in formulas of first-order logic. Now we will abandon this level of precision and we will use $<$ in both cases.

With this convention in mind, let $\text{Succ}(x, y)$ be the following first-order formula expressing that y is the successor of x.

$$x < y \wedge \forall z[x < z \implies (y < z \vee y = z)].$$

Let φ be the sentence

$$\forall x[\exists y \ (x < y) \implies \exists y \ \text{Succ}(x, y)].$$

Then, $<$ discretely orders A if and only if $(A, <) \models \varphi$.

Note on notation: For better readability, the sentence φ above could have been written as

$$\forall x[\exists y \ (x < y) \implies \exists z \ \text{Succ}(x, z)].$$

17

Writing φ this way emphasizes that the evaluation of y in $(x < y)$ may not be the same as the evaluation of z in $\text{Succ}(x, z)$, although it could be. Formally, both forms of φ have the same truth value in any structure. The same symbol can not be used for different free variables in a formula, but it can for the variables bound by quantifiers. Which form to use is only a question of aesthetics.

PROPOSITION 3.2. Each natural number is definable in $(\mathbb{N}, <)$.

PROOF. The proof is by induction. For the base case, let $\varphi_0(x)$ be $\neg \exists y(y < x)$. The only element satisfying $\varphi_0(x)$ in $(\mathbb{N}, <)$ is 0, so 0 is definable.

For the inductive step, assume that a formula $\varphi_n(x)$ defines n in $(\mathbb{N}, <)$, and let $\varphi_{n+1}(x)$ be

$$\forall z[\varphi_n(z) \implies \text{Succ}(z, x)].$$

The only element satisfying $\varphi_{n+1}(x)$ in $(\mathbb{N}, <)$ is $n + 1$, so $n + 1$ is definable and this finishes the proof. □

Notice the special form of $\varphi_{n+1}(x)$. It seems more natural to simply write $\text{Succ}(n, x)$ instead. The problem here is that we do not have a symbol for n in the language of $(\mathbb{N}, <)$. We do have a definition of n and this allows us to use the universal quantifier $\forall z$ to identify n. Later we will allow expressions such as $\text{Succ}(n, x)$, with the understanding that while they are not fully formal, they can be converted to legitimate first-order formulas using universal quantification.

We will see in the next chapter that no integer is definable in $(\mathbb{Z}, <)$ and no rational number is definable in $(\mathbb{Q}, <)$.

EXERCISE 3.3. Prove that each integer is definable in $(\mathbb{Z}, <, 0)$.

3.2. Defining Rationals

With a bit of algebra, one can show that the neutral elements of addition and multiplication are definable in every field. It is straightforward, that they are definable in the field of rational numbers, as well in the fields or real numbers and complex numbers and in the ring of integers. In those structures 0 is the only element satisfying the formula $x = x + x$ and 1 is the only element satisfying $\neg(x = x + x) \wedge x = x \cdot x$. For this section, let us call these formulas $\varphi_0(x)$ and $\varphi_1(x)$, respectively.

If an element a is definable in a structure $\mathfrak{M}$, then a constant symbol for it does not have to be included in the language of $\mathfrak{M}$, because to refer to it we can always use its definition. Nevertheless, such constants for some definable elements are often in the language just for convenience.

PROPOSITION 3.4. Each rational number is definable in the field $(\mathbb{Q}, +, \cdot)$

PROOF. Let $s(x, y)$ be

$$\forall z[\varphi_1(z) \implies y = x + z].$$

Now, as in in Proposition 3.2, using $s(x, y)$ instead of $\text{Succ}(x, y)$, we can show that each natural number is definable in $(\mathbb{Q}, +, \cdot)$. It follows that each negative integer is definable as well; for each $n \geq 0$, if $\varphi_n(x)$ defines n, then

$$\forall y \forall z[(\varphi_n(y) \wedge \varphi_0(z)) \implies x + y = z]$$

defines $-n$.

For integers m and n, let $\theta_{m,n}(x)$ be the formula

$$\forall y \forall z[(\varphi_n(y) \wedge \varphi_m(z)) \implies y \cdot x = z].$$

If $n \neq 0$, then $\dfrac{m}{n}$ is the only rational number p such that $\theta_{m,n}(p)$ holds in $(\mathbb{Q}, +\cdot)$. $\qquad\square$

EXERCISE 3.5. What do $\theta_{0,0}(x)$ and $\theta_{1,0}(x)$ define in $(\mathbb{Q}, +, \cdot)$?

Notice that the formulas introduced in the proof of Proposition 3.4 define all rational numbers not only in $(\mathbb{Q}, +, \cdot)$, but in any field that contains $\mathbb{Q}$, such as the field of real numbers $(\mathbb{R}, +, \cdot)$ and the field of complex numbers $(\mathbb{C}, +, \cdot)$.

3.3. Defining Order

The ordering of the natural numbers is defined in $(\mathbb{N}, +)$ by the formula

$$\neg(x = y) \wedge \exists z(x + z = y).$$

As we will see later, the ordering of $\mathbb{Z}$ is not definable in $(\mathbb{Z}, +)$, but now we will see that it is definable in $(\mathbb{Z}, +, \cdot)$. To show it, we need some bigger guns. Lagrange's four squares theorem states that every natural number is the sum of four squares. Hence,

$$\varphi_{\mathbb{N}}(x) := \exists x_1 \exists x_2 \exists x_3 \exists x_4 [x = x_1 \cdot x_1 + x_2 \cdot x_2 + x_3 \cdot x_3 + x_4 \cdot x_4]$$

defines $\mathbb{N}$ in $(\mathbb{Z}, +, \cdot)$. The ordering of $\mathbb{Z}$ can be now defined in $(\mathbb{Z}, +, \cdot)$ by

$$\neg(x = y) \wedge \exists z[\varphi_{\mathbb{N}}(z) \wedge (x + z = y)].$$

EXERCISE 3.6. For all m and n, $n \neq 0$, $\frac{m}{n} = \frac{nm}{n^2}$. Use this and Lagrange's four squares theorem to show that the ordering of $\mathbb{Q}$ is definable in $(\mathbb{Q}, +, \cdot)$.

EXERCISE 3.7. Show that the ordering is definable in the field of real numbers $(\mathbb{R}, +, \cdot)$.

3.4. Parametric Definability

Up to now, we have considered definability by formulas of the language of $\mathfrak{M}$. Those formulas can refer to elements that are definable, but not to arbitrary elements of M. However, in many applications of model theory, definability involving all elements is crucial. In Chapter 2, we saw how to formally interpret expressions such as $\mathfrak{M} \models \varphi(a_1, \ldots, a_n)$, where $a_1, \ldots, a_n$ are in M. We use this interpretation in the following definition.

DEFINITION 3.8. Let $\mathfrak{M}$ be a structure, and let X be a subset of M^n for some $n > 0$. We say that X is *parametrically definable* in $\mathfrak{M}$ if there are a formula $\varphi(x_1, \ldots, x_n, y_1, \ldots, y_m)$ and $b_1, \ldots, b_m$ in M such that

$$X = \{(a_1, \ldots, a_n) : \mathfrak{M} \models \varphi(a_1, \ldots, a_n, b_1, \ldots, b_m)\}.$$

Let $\mathfrak{M}$ be a structure, and let $X = \{b_1, \ldots, b_m\}$ be a finite subset of M. Then, X is defined in $\mathfrak{M}$ by the formula

$$\varphi_X(x) = [(x = b_1) \vee \cdots \vee (x = b_m)],$$

and the complement of X in M is defined by $\neg \varphi_X(x)$. This shows that all finite and all cofinite[1] subsets of M are parametrically definable and a similar argument applies to subsets of M^n for any n. For example, $\{(b_1, c_1), (b_2, c_2)\}$ is defined by

$$[(x = b_1) \wedge (y = c_1)] \vee [(x = b_2) \wedge (y = c_2)].$$

We say that a structure $\mathfrak{M}$ is *linearly ordered* if one of its relations linearly orders M. Elements of the domain of a linearly ordered structure are often called *points*.

Let $\mathfrak{M}$ be linearly ordered by $<$. An *open interval* of $\mathfrak{M}$ is a set of one of the following forms for arbitrary a and b in M,

$$\{x : \mathfrak{M} \models (x < a)\},$$
$$\{x : \mathfrak{M} \models (a < x)\},$$
$$\{x : \mathfrak{M} \models (a < x) \wedge (x < b)\}.$$

In linearly ordered structures, in addition to all finite and all cofinite sets, all open intervals are definable. This leads us to the following definition.

DEFINITION 3.9. A structure is *minimal* if every parametrically definable subset of its domain is either finite or cofinite. A linearly ordered structure is *o-minimal* if every parametrically definable subset of its domain is the union of finitely many open intervals and points.

All structures with finite domains are minimal. To show nontrivial examples of minimal and o-minimal structures, we need a method of proving that certain sets are not definable. One such method will be developed in the next chapter. Now we will only list some non-examples.

EXAMPLE 3.10. $(\mathbb{Z}, <)$ is not minimal. The set defined by $x < 0$ is neither finite nor cofinite.

EXAMPLE 3.11. $(\mathbb{Z}, +)$ is not minimal. The set of even numbers is defined by $\exists y (y + y = x)$. This also shows that $(\mathbb{Z}, +, <)$ is not o-minimal.

EXAMPLE 3.12. $(\mathbb{Z}, \cdot)$ is not minimal. The set of squares is defined by $\exists y (y \cdot y = x)$.

---

[1]A set $X \subseteq M$ is cofinite if $M \setminus X$ is finite.

EXAMPLE 3.13. Let $S = \{(x, y) : x, y \in \mathbb{R} \land y = \sin x\}$. Then $(\mathbb{R}, <, S)$ is not o-minimal, since the set $\{x : (x, 0) \in S\}$ is not the union of finitely many open intervals and points.

CHAPTER 4

Symmetry

This chapter is about isomorphisms and automorphisms, so a word about the title is necessary. In [**Pie17**], David Pierce argues that "Commensurability and symmetry have diverged from a common Greek origin. [...] Today we can precisely define the symmetry of a mathematical structure as the automorphism group of the structure, or as the isomorphism class of that group."

4.1. Isomorphisms

For each relation symbol $\mathcal{R}$ in the language of a structure $\mathfrak{M}$, $\mathcal{R}_{\mathfrak{M}}$ denotes the relation of $\mathfrak{M}$ that is the interpretation of $\mathcal{R}$.

DEFINITION 4.1. Let $\mathfrak{M}$ and $\mathfrak{N}$ be structures with the same language. Then, $\mathfrak{M}$ and $\mathfrak{N}$ are *isomorphic* if there is a one-to-one and onto function $f : M \longrightarrow N$ such that for each constant symbol c, $f(c_{\mathfrak{M}}) = c_{\mathfrak{N}}$ and for each n-ary relation symbol $\mathcal{R}$ and all $a_1, ..., a_n$ in M

$$(a_1, \ldots, a_n) \in \mathcal{R}_{\mathfrak{M}} \text{ iff } (f(a_1), \ldots, f(a_n)) \in \mathcal{R}_{\mathfrak{N}},$$

EXAMPLE 4.2. Let $\mathfrak{M}$ be $(\mathbb{N}, <)$, and let $\mathfrak{N}$ be $(E, <)$, where E is the set of even numbers. The function $f : \mathbb{N} \longrightarrow E$, defined by $f(n) = 2n$ is one-to-one, onto and it preserves the ordering; hence it is an isomorphism.

EXERCISE 4.3. Let $\mathfrak{M}$ be $(\mathbb{N}, <)$, let X be an infinite subset of $\mathbb{N}$, and let $\mathfrak{N}$ be $(X, <)$ with the ordering inherited from $\mathfrak{M}$. Define an isomorphism between $(\mathbb{N}, <)$ and $(X, <)$.

EXAMPLE 4.4. Let $\mathfrak{M}$ be $(\mathbb{N}, <)$, and let $\mathfrak{N}$ be $(\{(n, n) : n \in \mathbb{N}\}, <)$, with the ordering defined by $(m, m) < (n, n)$ if and only if $m < n$. It is clear that these structures are isomorphic. Their domains are different, but as ordered sets they are really the same structure.

In mathematics, we often identify structures up to isomorphism. In analyzing structure in ways that we will be discussing here, it is not important what the elements of the domain of a structures are made of, it only matters how they are related to one another by the relations of the structure.

Recall the group $\mathbb{Z}_2 = (\{0, 1\}, +, 0)$ defined right after Definition 1.7. Let $\mathfrak{G} = (\{e, a\}, \circ, e)$ be a two-element group. It is easy to see that the map $0 \mapsto e$, $1 \mapsto a$ is an isomorphism. So, up to isomorphism, there is only one two-element group and we can say that $\mathfrak{G}$ *is* $\mathbb{Z}_2$. In this sense, the group $(\{1, -1\}, \cdot, 1)$, where

· is the usual multiplication, is $\mathbb{Z}_2$, although in algebra it is usually referred to as the multiplicative group on 1 and -1, or the *sign group*.

The proof of the next important theorem is straightforward, but we will take a close look at the details. It is an example of a proof by induction on the complexity of formulas.

The proof uses the notion of *rank* of a formula.

DEFINITION 4.5. If φ is atomic, then $\text{rk}(\varphi) = 0$. For non-atomic formulas,

$$\text{rk}(\varphi \wedge \psi) = \max\{\text{rk}(\varphi), \text{rk}(\psi)\} + 1,$$

$$\text{rk}(\neg\varphi) = \text{rk}(\varphi) + 1,$$

$$\text{rk}(\exists x\varphi) = \text{rk}(\varphi) + 1.$$

A verification is needed that the rank function is well-defined. This follows from the unique reading lemma that says that each formula is either atomic, or is of one of the forms in Definition 4.5 for unique φ and ψ. For an interesting proof, see [**Man10**].

Recall that for $\bar{a} = (a_1, \ldots, a_n)$ and $f : M \longrightarrow N$, $f(\bar{a})$ is an abbreviation for $(f(a_1), \ldots, f(a_n))$.

THEOREM 4.6. *Let* $\mathfrak{M}$ *and* $\mathfrak{N}$ *be structures with the same language. If* $f : M \longrightarrow N$ *is an isomorphism, then for all* $n > 0$, *for all formulas* $\varphi(\bar{x})$ *with* n *free variables and for all* $\bar{a}$ *in* M^n

$$\mathfrak{M} \models \varphi(\bar{a}) \text{ iff } \mathfrak{N} \models \varphi(f(\bar{a})).$$

PROOF. The proof is by induction on the rank of formulas defined above. The base case for formulas of rank 0 is: for each n-ary relation symbol $\mathcal{R}$ and all $\bar{a}$ in M^n

$$\mathfrak{M} \models \mathcal{R}(\bar{a}) \text{ iff } \mathfrak{N} \models \mathcal{R}(f(\bar{a})),$$

and that is exactly what Definition 4.1 says.

The inductive assumption is: for all formulas $\varphi(\bar{x})$ of rank at most k, all $n > 0$ and all $\bar{a}$ in M^n

$$\mathfrak{M} \models \varphi(\bar{a}) \text{ iff } \mathfrak{N} \models \varphi(f(\bar{a})).$$

Let $\varphi(\bar{x})$ be of rank $k+1$. If $\varphi(\bar{x})$ is of the form $\psi(\bar{x}) \wedge \theta(\bar{x})$, then the ranks of $\psi(\bar{x})$ and $\theta(\bar{x})$ are at most k, and

$$\mathfrak{M} \models \varphi(\bar{a}) \iff \mathfrak{M} \models \psi(\bar{a}) \text{ and } \mathfrak{M} \models \theta(\bar{a}) \text{ (by Definition 2.6)}$$
$$\iff \mathfrak{N} \models \psi(f(\bar{a})) \text{ and } \mathfrak{N} \models \theta(f(\bar{a})) \text{ (by the inductive assumption)}$$
$$\iff \mathfrak{N} \models \varphi(f(\bar{a})) \text{ (by Definition 2.6)}$$

The case of $\neg\varphi$ is similar.

The last case to check is when φ be of the form $\exists x \psi(x, \bar{y})$. Then,

$$\mathfrak{M} \models \varphi(\bar{a}) \Longleftrightarrow \mathfrak{M} \models \psi(b, \bar{a}) \text{ for some } b \text{ in } M \quad \text{(by Definition 2.6)}$$
$$\Longleftrightarrow \mathfrak{N} \models \psi(f(b, \bar{a})) \text{ (by the inductive assumption)}$$
$$\Longleftrightarrow \mathfrak{N} \models \varphi(f(\bar{a})) \text{ (by Definition 2.6)}$$

and this finishes the proof. □

To check whether a given function f is an isomorphism, it is enough to verify the condition in Definition 4.1 that involves only atomic formulas. To show that f is not an isomorphism, by Theorem 4.6, it is enough to find one first-order formula φ that violates the conclusion of the theorem for some $\bar{a}$.

EXERCISE 4.7. $(\mathbb{N}, <)$ is not isomorphic to $(\mathbb{Z}, <)$. Prove it using Theorem 4.6.

For an ordered set $(A, <)$ let the dual ordering $<^*$ be defined by $a <^* b$ if and only if $b < a$.

EXERCISE 4.8. Show that $(\mathbb{N}, <)$ and $(\mathbb{N}, <^*)$ are not isomorphic, but $(\mathbb{Z}, <)$ and $(\mathbb{Z}, <^*)$ are.

EXERCISE 4.9. Show that $(Z, +)$ is not isomorphic to $(\mathbb{Z}, \cdot)$.

EXERCISE 4.10. $(\mathbb{Z} \times \mathbb{Z}, +)$ is the group whose elements are ordered pairs of integers and $+$ is defined by $(k, l) + (m, n) = (k + m, l + n)$. Show that $(\mathbb{Z} \times \mathbb{Z}, +)$ is not isomorphic to $(\mathbb{Z}, +)$.

EXERCISE 4.11. $(\mathbb{Z} \oplus \mathbb{Z}, <)$ is a linearly ordered set consisting of two copies of $(\mathbb{Z}, <)$ one on top of the other,i.e., all elements of the first $(\mathbb{Z}, <)$ are smaller than all elements of the second and within each copy the ordering is the usual ordering on $\mathbb{Z}$. Prove that $(\mathbb{Z} \oplus \mathbb{Z}, <)$ is not isomorphic to $(\mathbb{Z}, <)$.

4.2. Automorphisms

In Definition 4.1, nothing prevents $\mathfrak{M}$ and $\mathfrak{N}$ from being the same. In this case an isomorphism $f : M \longrightarrow M$ is called an *automorphism* of $\mathfrak{M}$. Any structure has at least one automorphism, namely the identity function $id(x) = x$. We call it the *trivial* automorphism. Let $\text{Aut}(\mathfrak{M})$ be the set of all automorphisms of $\mathfrak{M}$.

EXERCISE 4.12. For f and g in $\text{Aut}(\mathfrak{M})$, $f \circ g$ is the composition of f and g,i.e., for all a in M, $(f \circ g)(a) = f(g(a))$. Show that $(\text{Aut}(\mathfrak{M}), \circ, id)$ is a group.

We will refer to $\text{Aut}(\mathfrak{M})$ as the *automorphism group* of $\mathfrak{M}$.

If $f : A \longrightarrow B$ is a function and X is a subset of A, then the *image* of X under f is $f(X) = \{f(x) : x \in X\}$.

The following theorem is an excellent tool for detecting undefinable sets.

THEOREM 4.13. *Let X be a set that is definable in $\mathfrak{M}$ by a formula with parameters $\bar{b} = b_1, \ldots, b_n$, and let f be an automorphism of $\mathfrak{M}$ such that $f(\bar{b}) = \bar{b}$. Then $f(X) = X$.*

PROOF. Suppose that $\varphi(\bar{x}, \bar{b})$ defines $X \subseteq M^n$ in $\mathfrak{M}$. Let $\bar{a}$ be an element of X. Then, $\mathfrak{M} \models \varphi(\bar{a}, \bar{b})$. Since $f(\bar{b}) = \bar{b}$, by Theorem 4.6, $\mathfrak{M} \models \varphi(f(\bar{a}), \bar{b})$; hence $f(\bar{a}) \in X$. This shows that $f(X) \subseteq X$. The same argument applied to $\neg\varphi(\bar{x}, \bar{b})$, shows that $f(M^n \setminus X) \subseteq M^n \setminus X$. Because f is onto, $f(X) = X$. □

COROLLARY 4.14. *If an element a is definable in $\mathfrak{M}$, then $f(a) = a$, for all f in $\mathrm{Aut}(\mathfrak{M})$.*

EXERCISE 4.15. Let $(G, \circ, e)$ be a group. Show that e is definable in $(G, \circ)$.

EXERCISE 4.16. Let $+_\mathfrak{G}$ be addition defined on the set $\{0, 1, 2\}$ as follows: 0 is the neutral element and $1 +_\mathfrak{G} 1 = 2$, $1 +_\mathfrak{G} 2 = 2 +_\mathfrak{G} 1 = 0$, and $2 +_\mathfrak{G} 2 = 1$. Show that $\mathfrak{G} = (\{0, 1, 2\}, +_\mathfrak{G}, 0)$ is a group and prove that 0 is the only definable element in it, by showing that f defined by $f(0) = 0$, $f(1) = 2$ and $f(2) = 1$ is an automorphism of $\mathfrak{G}$.

A structure is *rigid* if it has no nontrivial automorphisms. It follows from Corollary 4.14 that any structure all of whose elements are definable is rigid.

The trivial group consisting of the neutral elements and $\mathbb{Z}_2$ are rigid. For another example, let us consider $(\{0, 1, 2\}, +, \cdot)$, where $+$ is the same as $+_\mathfrak{G}$ defined in the exercise above and $(\{1, 2\}, \cdot)$ is the multiplicative version of the group $\mathbb{Z}_2$, i.e., 1 is the neutral element of the group and $2 \cdot 2 = 1$. It is not difficult to check that $(\{0, 1, 2\}, +, \cdot)$ is a field. It can be shown that, up to isomorphism, it is the only field with three elements. In algebra it is known as $\mathbb{F}_3$. Every element of $\mathbb{F}_3$ is definable: 0 is the neutral element of addition, 1 is the neutral element of multiplication and 2 is the only element of the field that is neither 0 nor 1. Hence $\mathbb{F}_3$ is rigid.

Because every natural number is definable in $(N, <)$, that structure is rigid. In contrast, $(\mathbb{Z}, <)$ has infinitely many automorphisms. For each integer n, the function $f_n : \mathbb{Z} \longrightarrow \mathbb{Z}$ defined by $f_n(x) = x + n$ preserves the ordering; hence it is an automorphism.

EXERCISE 4.17. Show that if f is an automorphism of $(\mathbb{Z}, <)$, then f is one of the functions f_n defined above.

PROPOSITION 4.18. *If for all a and b in M there is an automorphism f of $\mathfrak{M}$ such that $f(a) = b$, then the only subsets of the domain that are definable in $\mathfrak{M}$ are M and the empty set.*

PROOF. If the set defined by $\varphi(x)$ is nonempty, then there is an a such that $\mathfrak{M} \models \varphi(a)$. Then, for any automorphism f, $\mathfrak{M} \models \varphi(f(a))$. By the assumption it follows that $\mathfrak{M} \models \varphi(b)$, for any b in M. □

Because for all integers a and b there is an automorphism f such that $f(a) = b$, by Proposition 4.18 the only sets of integers that are definable are $\mathbb{Z}$ and the empty set.

The only automorphism of $(\mathbb{N}, <)$ is identity; $(\mathbb{Z}, <)$ has countably many automorphisms and we know exactly what they are. There is more that one can say about automorphisms of $(\mathbb{Q}, <)$ and we will do it later. For now, the reader who is familiar with basic set theory should try the following exercise. By $2^{\aleph_0}$ we denote the cardinality of the set of all sets of natural numbers.

EXERCISE 4.19. Show that $\mathrm{Aut}((\mathbb{Q}, <))$ is of cardinality $2^{\aleph_0}$. HINT: Think of piecewise linear increasing functions.

We will finish this chapter with several results showing that some relations are not first-order definable in some of the structures that we discussed so far. The common feature of those results is the use of automorphisms.

4.3. Addition is not Definable in $(\mathbb{Z}, <)$

In the proof of Proposition 3.2, we used the formula $\mathrm{Succ}(x, u)$ that defines the successor relation in ordered structures. We will use it again.

DEFINITION 4.20. Let $\mathrm{Succ}_0(x, y)$ be $x = y$ and inductively define $\mathrm{Succ}_{n+1}(x, y)$ to be

$$\forall z [\mathrm{Succ}_n(x, z) \implies \mathrm{Succ}(z, y)].$$

For each non-negative integer n, the relation $x + n = y$ is defined in $(\mathbb{Z}, <)$ by $\mathrm{Succ}_n(x, y)$. One can also write a similar formula defining $x + n = y$ for negative n.

EXERCISE 4.21. For each negative integer n, define a formula that defines in $(\mathbb{Z}, <)$ the relation $x + n = y$.

While for each integer n, the relation $x + n = y$ is definable in $(\mathbb{Z}, <)$, the relation $x + z = y$ is not. To see why, suppose there is a formula $\varphi(x, z, y)$ of the language of $(\mathbb{Z}, <)$ such that for all integers a, b and c

$$(\mathbb{Z}, <) \models \varphi(a, b, c) \text{ iff } a + b = c.$$

Consider the automorphism $f(x) = x + 1$. Since $\varphi(x, z, y)$ defines addition, we have

$$(\mathbb{Z}, <) \models \varphi(0, 0, 0).$$

Because f is an automorphism and $f(0) = 1$, it follows that

$$(\mathbb{Z}, <) \models \varphi(1, 1, 1),$$

and this is a contradiction.

4.4. Ordering is not Definable in $(\mathbb{Z}, +)$

We have seen in Chapter 3 that the ordering of the integers is definable in $(\mathbb{Z}, +, \cdot)$. In the definition, both addition and multiplication were used in essential ways. That was necessary and that is because, as we will now show, the ordering is not definable in $(\mathbb{Z}, +)$.

$(\mathbb{Z}, +)$ has fewer automorphisms than $(\mathbb{Z}, <)$. In fact, it has only one non-trivial automorphism, but fortunately it is the one that works.

Suppose that $f : \mathbb{Z} \longrightarrow \mathbb{Z}$ is an automorphism of $(\mathbb{Z}, +)$. Because 0 is definable in $(\mathbb{Z}, +)$, $f(0) = 0$.

Suppose $f(1) = a$. Then $f(2) = f(1 + 1) = f(1) + f(1) = 2a$, $f(3) = f(1 + 1 + 1) = f(1) + f(1) + f(1) = 3a$, and so on. For each positive n, $f(n) = na$. Also, $f(n) + f(-n) = f(n + (-n)) = f(0) = 0$; hence, for positive n, $f(-n) = -f(n) = -na$. This shows that every automorphism of $(\mathbb{Z}, +)$ must be of the form $f_a(x) = ax$, for some integer $a \neq 0$.

If $|a| > 1$, then f_a is not onto, so it is disqualified, f_1 is the identity and we are left with f_{-1}. We have assumed that f_{-1} is a automorphism, but now we have to make sure that it is, but that is immediate because for all m, n, and k, $m + n = k$ if and only if $(-m) + (-n) = (-k)$.

Suppose now that there is a formula $\varphi(x, y)$ of the language of $(\mathbb{Z}, +)$ such that for all a and b,

$$(\mathbb{Z}, +) \models \varphi(a, b) \text{ iff } a < b.$$

Then $(\mathbb{Z}, +) \models \varphi(0, 1)$ and because $f(0) = 0$ and $f(1) = -1$, it follows that $(\mathbb{Z}, +) \models \varphi(0, -1)$, implying that $0 < -1$, contradiction.

Notice that because the automorphism $f(x) = -x$ moves all integers except 0, no integer other than 0 is definable in $(\mathbb{Z}, +)$.

Recall that when we say that a relation is definable, we mean definability without parameters. The arguments we used above cannot show that addition is not parametrically definable in $(\mathbb{Z}, <)$, nor that the ordering is not parametrically definable in $(\mathbb{Z}, +)$ because for that we would need automorphisms that fix the parameters of any potential definition. As we have seen, there are no such automorphisms. Both these undefinability results hold for parametric definability as well. Theorem 8.10 implies that the set of even numbers cannot be parametrically defined in $(\mathbb{Z}, <)$; hence $<$ cannot be parametrically defined. The proof that $<$ is not parametrically defined in $(\mathbb{Z}, +)$ follows from a more advanced general result on definability in Abelian groups.

4.5. Addition is not Definable in $(\mathbb{N}, \cdot)$

The fundamental theorem of arithmetic says that every natural number is the product of powers of its prime divisors and that the sequence of prime powers in the product is unique up to permutation of factors. In particular, every natural number can be written as $2^m 3^n a$, where a is divisible by neither 2 nor 3. Let $f : \mathbb{N} \longrightarrow \mathbb{N}$ be defined by $f(2^m 3^n a) = 2^n 3^m a$.

EXERCISE 4.22. Find $f(2)$, $f(3)$, $f(4)$, $f(5)$, and $f(12)$.

We will show that f is a automorphism of $(\mathbb{N}, \cdot)$. Let's compute

$$f((2^m 3^n a) \cdot (2^k 3^l b)) = f(2^{m+k} 3^{n+l} ab)$$
$$= 2^{n+l} 3^{m+k} ab$$
$$= 2^n 3^m a \cdot 2^l 3^k b$$
$$= f(2^m 3^n a) \cdot f(2^k 3^l b).$$

This is interesting on its own, but there are further consequences. Because f permutes 2 and 3, it does not preserve the ordering of $\mathbb{N}$. It follows that $<$ is not definable in $(\mathbb{N}, \cdot)$ and, because $<$ is definable in $(\mathbb{N}, +)$, we have also proved that addition is not definable in $(\mathbb{N}, \cdot)$.

Instead of 2 and 3 in the argument above we could have used any other pair of prime numbers. This can be used to show that $<$ is not parametrically definable in $(\mathbb{N}, \cdot)$. For every formula with parameters, we can find an automorphism of $(\mathbb{N}, <)$ that fixes the parameters and permutes two primes p and q. Then $(\mathbb{N}, \cdot) \models \varphi(p, q, \bar{b})$ if and only if $(\mathbb{N}, \cdot) \models \varphi(q, p, \bar{b})$; hence $\varphi(x, y, \bar{b})$ is not a definition of the ordering of $\mathbb{N}$.

EXERCISE 4.23. Prove that $\mathrm{Aut}((\mathbb{N}, \cdot))$ is of cardinality $2^{\aleph_0}$.

4.6. Multiplication is not Definable in $(\mathbb{N}, +)$

There are several ways in which the result in the title of this section can be proved, but none of them can be presented here with all details. One of the proofs uses a characterization of sets of natural numbers that are definable in $(\mathbb{N}, +)$, due to Ginsburg and Spanier [**GS66**].

DEFINITION 4.24. A set $X \subseteq \mathbb{N}$ is *ultimately periodic* if there are natural numbers k and p such that for all $m > k$, m is in X if and only if $m + p$ is in X.

Ginsburg and Spanier proved that sets of natural numbers that are definable in $(\mathbb{N}, +)$ are exactly the ultimately periodic sets. See [**Smo91**] for a proof and a discussion of the result. The set of squares is definable in $(\mathbb{N}, \cdot)$ and because it is not ultimately periodic, multiplication cannot be defined in $(\mathbb{N}, +)$.

4.7. Definability in $(\mathbb{Q}, +)$

This section is a list of exercises for the reader.

EXERCISE 4.25. Prove that for each nonzero rational number p, $f_p : \mathbb{Q} \longrightarrow \mathbb{Q}$ defined by $f(x) = px$ is an automorphism of $(\mathbb{Q}, +)$.

EXERCISE 4.26. Let $f : \mathbb{Q} \longrightarrow \mathbb{Q}$ be an automorphism of $(\mathbb{Q}, +)$. Prove that f is one of the automorphisms defined in the previous exercise.

EXERCISE 4.27. Prove that 0 is the only rational number that is definable in $(\mathbb{Q}, +)$.

EXERCISE 4.28. Prove that the ordering of $\mathbb{Q}$ is not definable in $(\mathbb{Q}, +)$.

EXERCISE 4.29. Prove that multiplication is not definable in $(\mathbb{Q}, +, <)$.

4.8. Orderable Fields

An ordered structure $(F, +, \cdot, <)$ is an *ordered field* if $(F, +, \cdot)$ is a field and the ordering satisfies the following axioms:

(1) $\forall x, y, z[x < y \implies x + z < y + z]$.
(2) $\forall x, y, z[(x < y \wedge 0 < z) \implies x \cdot z < y \cdot z]$.

A field is *orderable* if it can be expanded to an ordered field. In Section 3.3 we saw that the usual ordering of numbers is definable in $(\mathbb{Z}, +, \cdot)$ and the reader was asked to verify the same for the fields $(\mathbb{Q}, +, \cdot)$ and $(\mathbb{R}, +, \cdot)$. Hence, the fields of rational and real numbers are orderable and one can check that their usual orderings are the only orderings that satisfy the ordered field axioms. It is also not difficult to prove that in ordered fields all squares must be positive,i.e., they are greater than 0 and this implies that the field of complex numbers is not orderable.

We will finish this chapter with an example of field that is orderable, but has no definable ordering.

Let $\mathbb{Q}(\sqrt{2})$ be the set of all real numbers of the form $a + b\sqrt{2}$, where a and b are rational. It is easy to check that $\mathbb{Q}(\sqrt{2})$ is closed under addition and multiplication and that the inverse of every nonzero number in $\mathbb{Q}(\sqrt{2})$ is also in $\mathbb{Q}(\sqrt{2})$. Hence $\mathbb{Q}(\sqrt{2})$ is a subfield of the field of real numbers and it is an ordered field with the ordering $<$ inherited from $\mathbb{R}$. We will show that that ordering is not definable in $(\mathbb{Q}(\sqrt{2}), +, \cdot)$.

Let $f : \mathbb{Q}(\sqrt{2}) \longrightarrow \mathbb{Q}(\sqrt{2})$ be defined by $f(a + b\sqrt{2}) = a - b\sqrt{2}$. In particular, $f(\sqrt{2}) = -\sqrt{2}$. By a direct calculation one can show that f is an automorphism of $(\mathbb{Q}(\sqrt{2}), +, \cdot)$. This implies that $<$ is not definable in $(\mathbb{Q}(\sqrt{2}), +, \cdot)$. If it were, that would mean that $-\sqrt{2} < \sqrt{2}$ and $\sqrt{2} < -\sqrt{2}$, giving us a contradiction.

EXERCISE 4.30. Verify all claims made in the two paragraphs above.

The ordering of the reals is definable in the real field because every positive number has a square root and the ordering of the rationals is definable because every positive rational number is the sum of the squares of four rational numbers. Because

$$(a + b\sqrt{2})^2 = a^2 + 2b^2 + 2ab\sqrt{2},$$

no such number in $\mathbb{Q}(\sqrt{2})$ can be equal to $\sqrt{2}$. This also shows that $\sqrt{2}$ can not be equal to the sum of any number of squares in $\mathbb{Q}(\sqrt{2})$.

CHAPTER 5

Theories and Types

Despite the preamble to the chapter on definability, to know a structure it is not enough to know its definable sets, one also needs to know the types of its elements and that also includes its complete theory. These are two central notions that will be defined in this chapter.

5.1. Theories

Group axioms, field axioms, Peano's axioms for arithmetic, Zermelo-Fraenkel set theory are examples of specific first-order theories but the definition of theory below does not restrict what counts as a theory to such examples. It says that a first-order theory is *any* set of sentences in a given first-order language. While in practice there are only some specific theories that we are interested in, the door is open for many more, because, and—as you will see—this is very useful. In these lectures, instead of "first-order theory" we will just say "theory."

DEFINITION 5.1. A *theory* T is a set of first-order sentences in a fixed language $\mathcal{L}_T$. A structure $\mathfrak{M}$ is a *model* of a theory T if for every φ in T, $\mathfrak{M} \models \varphi$. A theory is *consistent* if it has a model. For a structure $\mathfrak{M}$ *the theory of* $\mathfrak{M}$ is the set of all sentences of the language of $\mathfrak{M}$ that are true in $\mathfrak{M}$ and is denoted $\mathrm{Th}(\mathfrak{M})$.

Since its beginnings in the 1950's, model theory could be described as an effort to answer the following two general questions: Given a consistent theory T, what can we say about the class of all models of T? Given a structure $\mathfrak{M}$, what can we say about $\mathrm{Th}(\mathfrak{M})$?

A representative example of a problem of the first kind that generated much research is the Vaught conjecture that if a theory in a countable language has uncountably many countable models, then it has $2^{\aleph_0}$ countable models. The continuum hypothesis, abbreviated CH, says that the smallest uncountable cardinal number is $2^{\aleph_0}$, so if CH holds, the Vaught conjecture holds as well. However, if CH fails, the Vaught conjecture could still hold and that would mean that one cannot find counterexamples to CH by counting the numbers of countable models of first-order theories. The Vaught conjecture is open.

An example of a problem of the second kind is Tarski's question whether the theory of the field of the real numbers with the exponential function is decidable, i.e., whether there is an algorithm to decide whether a given sentence in the language of this structure is true or false. This question is also open, although it was shown by Angus Macintyre and Alex Wilkie [**MW96**] that it has a positive solution assuming a conjecture in transcendental number theory known as Schanuel's conjecture.

Because we consider any set of sentences a theory, $\{\varphi, \neg\varphi\}$ and $\{\exists x \neg(x = x)\}$ are theories and they are clearly inconsistent. The compactness theorem, which is the topic of Chapter 7, says that any theory that does not include such obvious inconsistencies, in a precise sense to be defined, is in fact consistent, i.e., it has a model. This turns out to be of great importance.

5.2. Types

In a language with function symbols, equations are formulas of a particular form. We can think of arbitrary formulas as generalized equations. Recall that for a set M, $M^{<\omega}$ is the set of all finite sequences of the elements of M. For a structure $\mathfrak{M}$, the set defined by a formula $\varphi(\bar{x})$, can be thought as a set of those $\bar{a}$ in $M^{<\omega}$ that are "solutions" to this formula. There is also a reverse direction; for a tuple $\bar{a}$ in $M^{<\omega}$, we can ask: what are the formulas $\varphi(\bar{x})$ for which $\bar{a}$ is a solution. So this is like asking: here is a number, what are the equations that this number is a solution of? This idea is captured in the notion of type.

DEFINITION 5.2. Let $\mathfrak{M}$ be a structure, and let $\bar{a}$ be an n-tuple in M^n. The *type* of $\bar{a}$, denoted $\text{tp}^{\mathfrak{M}}(\bar{a})$, is the set $\{\varphi(\bar{x}) : \mathfrak{M} \models \varphi(\bar{a})\}$.

The superscript in $\text{tp}^{\mathfrak{M}}(\bar{a})$ will sometimes be dropped if it is clear from the context what $\mathfrak{M}$ is.

Note on notation: By definition, the type of a tuple $\bar{a} = a_1, \ldots, a_n$ is a collection of of formulas with n free variables. However, if φ is a sentence of the language of the structure, we can let φ^* be $\varphi \wedge (x_1 = x_1) \wedge \cdots (x_n = x_n)$, and then φ^* is in $\text{tp}(\bar{a})$ if and only if φ is true in $\mathfrak{M}$. We can deal similarly with any formula with fewer than n free variables. This means that the type of an n-tuple codes in a natural way $\text{Th}(\mathfrak{M})$ and the k-types of all subtuples of $\bar{a}$. In this sense, we can say $\text{tp}(\bar{a})$ includes $\text{Th}(\mathfrak{M})$. Another, perhaps simpler way to get $\text{Th}(\mathfrak{M}) \subseteq \text{tp}(\bar{a})$, is by declaring that each formula of arity k is automatically considered of arity n for all $n > k$ (the same way as we can think of a polynomial of degree k as a degenerate polynomial of degree n.)

Although it is easy to say what the type of an element or a tuple is, it is often much harder to classify or just identify some important types in a given structure. In this chapter we will see how it is done for basic ordered sets but even for some of those structures we will have to wait for a complete analysis until more powerful model-theoretic techniques are developed.

In some structures, each element has its own unique type. This seems to be an interesting case, but in fact it is just the opposite. Much work in model theory is done in structures with lots of elements and tuples sharing the same type, which are moreover such that if $\text{tp}^{\mathfrak{M}}(\bar{a}) = \text{tp}^{\mathfrak{M}}(\bar{b})$, then there is an automorphism of $\mathfrak{M}$ that takes $\bar{a}$ to $\bar{b}$. In such situations, there is much relevant information about $\mathfrak{M}$ that can be recovered from the automorphism group of $\mathfrak{M}$ and this kind of research has become a discipline on its own. This is beyond of what can be presented in introductory lectures, but in later chapters we will discuss some central ideas involving types—saturation and indiscernibility—that are crucial in the study of automorphism groups of first order structures. The book [**KM94**] is a collection of articles on the subject with preliminary chapters written by Richard Kaye and Dugald Macpherson, that I recommend for further reading.

The next basic proposition follows directly from definitions.

PROPOSITION 5.3. *If $\mathfrak{M}$ and $\mathfrak{N}$ are isomorphic, then for every isomorphism $f : M \longrightarrow N$ and all $\bar{a}$ in $\mathfrak{M}$, $\mathrm{tp}^{\mathfrak{M}}(\bar{a}) = \mathrm{tp}^{\mathfrak{N}}(f(\bar{a}))$.*

The reader is encouraged to write a proof as an exercise, but it is a kind of result that, for more advanced readers, should be so obvious that a proof is hardly needed. If $f : M \longrightarrow N$ is an isomorphism and $f(\bar{a}) = \bar{b}$, then the structures $(\mathfrak{M}, \bar{a})$ and $(\mathfrak{N}, \bar{b})$[1] are isomorphic; hence $\bar{a}$ "looks" in $\mathfrak{M}$ exactly as $\bar{b}$ "looks" in $\mathfrak{N}$, in particular, they should have the same first-order properties and this means that they have the same types. (That of course, was not a proof).

For each $\bar{a}$, $\mathrm{tp}(\bar{a})$ is an infinite set; for each formula $\varphi(\bar{x})$ of the language of $\mathfrak{M}$, either $\varphi(\bar{x})$ or $\neg\varphi(\bar{x})$ is in $\mathrm{tp}(\bar{a})$ (but clearly not both). It can happen though, that there is a single formula $\varphi(\bar{x})$ that determines the whole $\mathrm{tp}(\bar{a})$ as in the following definition.

DEFINITION 5.4. *Let $\bar{a}$ be in $M^{<\omega}$. If there is a formula $\varphi(\bar{x}) \in \mathrm{tp}^{\mathfrak{M}}(\bar{a})$ such that for all $\bar{b}$, if $\mathfrak{M} \models \varphi(\bar{b})$, then $\mathrm{tp}^{\mathfrak{M}}(\bar{b}) = \mathrm{tp}^{\mathfrak{M}}(\bar{a})$, then we say that $\mathrm{tp}^{\mathfrak{M}}(\bar{a})$ is isolated and that $\varphi(\bar{x})$ isolates the type.* Isolated types are also called *principal*.

EXERCISE 5.5. Show that if $\varphi(\bar{x})$ isolates $\mathrm{tp}^{\mathfrak{M}}(\bar{a})$, then $\varphi(\bar{x})$ is in $\mathrm{tp}^{\mathfrak{M}}(\bar{a})$.

EXERCISE 5.6. Show that if a is definable in $\mathfrak{M}$, then $\mathrm{tp}^{\mathfrak{M}}(a)$ is isolated.

So far, we have talked about types of elements in a given structure, but it also makes sense to consider types as just collections of formulas in a given language. The following definition introduces a mild restriction and some more terminology.

DEFINITION 5.7. *For each $n > 0$, an n-type is any collection of first-order formulas with at most n free variables in a fixed language. An n-type $p(\bar{x})$ is complete, if for each formula $\varphi(\bar{x})$ with at most n free variables, either $\varphi(\bar{x})$ or $\neg\varphi(\bar{x})$ is in $p(\bar{x})$. An n-type $p(\bar{x})$ is realized in a structure $\mathfrak{M}$ if for some $\bar{a}$ in M^n, $p(\bar{x}) \subseteq \mathrm{tp}^{\mathfrak{M}}(\bar{a})$. A type is an n-type for some n.* We say that a type is *consistent* if it is realized in some structure.

---

[1]The note following Theorem 6.4 explains this notation.

Definition 5.4 can be generalized to types as defined above as follows. If $p(\bar{x})$ is a type and T is a theory in the same language, then $p(\bar{x})$ is isolated in T if there is a formula $\psi(\bar{x})$ such that, for all models $\mathfrak{M}$ of T and all $\varphi(\bar{x})$ in $p(\bar{x})$, $\mathfrak{M} \models \forall \bar{x}[\psi(\bar{x}) \implies \varphi(\bar{x})]$. If the conclusion of the previous sentence holds for a given model $\mathfrak{M}$, then we say that $p(\bar{x})$ is isolated in $\mathfrak{M}$.

EXERCISE 5.8. Assume that for every sentence φ of the language of a theory T, either $\varphi \in T$ or $\neg \varphi \in T$. Prove that if $p(\bar{x})$ is an isolated type in one model of T, then it is isolated in all models of T.

5.3. Types in Ordered Structures

In any linearly ordered structure, if $a < b$ and $c > d$, then the types of (a, b) and (c, d) are different. The formula $x < y$ is in $\text{tp}((a, b))$ and the formula $\neg(x < y)$ is in $\text{tp}((c, d))$. In general, the type of an n-tuple depends on how the elements in it are ordered with respect to one another. Hence, for each n, each ordered structure realizes at least $n!$ different complete n-types.

To classify all types realized in an ordered structure, it is enough to consider tuples $\bar{a} = (a_1, \ldots, a_n)$ such that $a_1 < \cdots < a_n$. To see why, let us take a look at an example.

Let $\bar{a} = (a, b, c)$, with $a < b < c$, and let $\bar{b} = (b, a, c)$. For each formula $\varphi(x_1, x_2, x_3)$ let $\varphi^*(x_1, x_2, x_3)$ be $\varphi(x_2, x_1, x_3)$. Then $\varphi(\bar{x})$ is in $\text{tp}(\bar{a})$ if and only if $\varphi^*(\bar{x})$ is in $\text{tp}(\bar{b})$.

For a complete classification of all types, we also need to consider types of tuples with repeating elements. This can be easily done once the classification of types of tuples with distinct elements is done. This is left for the reader as an exercise.

5.3.1. Types in $(\mathbb{Z}, <)$. Because every natural number is definable in $(\mathbb{N}, <)$, each number has its own unique isolated type. Then, each tuple also has its own unique isolated type that is determined by the types of the elements in the tuple. The picture changes dramatically when we move to $(\mathbb{Z}, <)$.

In the previous chapter, we saw that for each pair of integers a and b, there is an automorphism f of $(\mathbb{Z}, <)$ such that $f(a) = b$. Hence, by Proposition 5.3, $\text{tp}(a) = \text{tp}(b)$. All integers share the same type in $(\mathbb{Z}, <)$ and that type is isolated by the formula $x = x$. We can say more about types of ordered pairs.

The formulas $\text{Succ}_n(x, y)$ from Definition 4.20 expresses that y is the n-th successor of x and all those formulas are in the language of $(\mathbb{Z}, <)$. It follows that if

$$(\mathbb{Z}, <) \models [\text{Succ}_i(a, b) \land \text{Succ}_j(c, d)],$$

and $i \neq j$, then $\text{tp}((a, b)) \neq \text{tp}((c, d))$. This gives us infinitely many types of pairs. All those types are isolated and each is isolated by $\text{Succ}_n(x, y)$, for some n. This is the case because if

$$(\mathbb{Z}, <) \models [\text{Succ}_i(a, b) \land \text{Succ}_i(c, d)],$$

and f is an automorphism defined by $f(x) = x + (c - a)$, then $f(a, b) = (f(a), f(b)) = (c, d)$. Hence $\mathrm{tp}((a, b)) = \mathrm{tp}((c, d))$ by Proposition 5.3.

Similarly, for any increasing tuple of integers $a_1 < \cdots < a_n$, the type of the tuple is determined by the distances between consecutive elements; hence, for each n there are infinitely many types of such tuples and all those types are isolated.

EXERCISE 5.9. Verify both claims in the last sentence above.

5.3.2. Types in $(\mathbb{Q}, <)$. While there are no nontrivial automorphisms of $(\mathbb{N}, <)$ and there are only countably many automorphisms of $(\mathbb{Z}, <)$, all of a very specific kind, $(\mathbb{Q}, <)$ has continuum many automorphisms. The reader was invited to prove it in Exercise 4.19.

As was the case of $(\mathbb{Z}, <)$, it is easy to see that the only sets of rational numbers that are definable in $(\mathbb{Q}, <)$ are $\mathbb{Q}$ and the empty set. Now, using automorphisms of $(\mathbb{Q}, <)$, we will be able to describe all parametrically definable sets and all types realized in $(\mathbb{Q}, <)$. We will begin with types.

If $a_1 < \cdots < a_n$ and $b_1 < \cdots < b_n$ are two increasing tuples of rational numbers, then there is an automorphism f of $(\mathbb{Q}, <)$ such that $f(a_i) = b_i$, for all $i = 1, \ldots, n$. To see that, draw a coordinate system. Mark the a_i's on the horizontal axis and the b_i's on the vertical. Plot all (a_i, b_i) and connect consecutive points with straight line segments. Then extend what you drew in both directions drawing half lines with slope 1. The picture shows the graph of an increasing function $f : \mathbb{Q} \longrightarrow \mathbb{Q}$, which is the required automorphism. One has to check that for each rational p, $f(p)$ is also rational. It is so because f is piecewise linear and each linear segment of it has a rational slope.

The argument above serves as a proof of the following proposition.

PROPOSITION 5.10. For each $n > 0$, there are exactly $n!$ complete n-types of n-tuples with distinct elements that are realized in $(\mathbb{Q}, <)$.

Next we will prove an important result on the uniqueness of the ordering of the rationals. It was first proved by Georg Cantor in 1885.

THEOREM 5.11. *Every countable densely linearly ordered set without end points is isomorphic to* $(\mathbb{Q}, <)$.[2]

PROOF. Let $(D, <)$ be countable densely linearly ordered set without end points. Let $\{q_i : i \in \mathbb{N}\}$ be an enumeration of $\mathbb{Q}$ and let $\{d_i : i \in \mathbb{N}\}$ be an enumeration of D. For each $n > 0$ we will define p_n in $\mathbb{Q}$ and e_n in D by induction. The inductive assumption is that for all i, $j \leq n$

$$p_i < p_j \text{ iff } e_i < e_j.$$

Let $p_0 = q_0$ and $e_0 = d_0$. The inductive assumption for $n = 0$ holds vacuously.

Assume that we have $P = \{p_i : i \leq n\}$ and $E = \{e_i : i \leq n\}$ for which the inductive assumption holds.

---

[2]A set X is countable is there is a one-to-one and onto function $f : \mathbb{N} \longrightarrow X$.

If n is even, we let $e_{n+1} = d_j$ for the smallest j such that d_j is not in E. If e_{n+1} is between two consecutive elements of E, say e_i and e_j, then, using density of the ordering, we let p_{n+1} be any rational number between p_i and p_j. If e_{n+1} is either below or above all elements in E, then we let p_{n+1} be any rational number that is either below or above all numbers in P, respectively. The inductive assumption is now satisfied for all $i, j \leq n+1$.

If n is odd, then the procedure is the same, but we begin with finding the smallest j such that q_j is not in P, and let it be p_{n+1} and then find e_{n+1} in D so that the inductive assumption is that for all $i, j \leq n+1$.

Now $\mathbb{Q} = \{p_i : i \in \mathbb{N}\}$ and $D = \{e_i : i \in \mathbb{N}\}$, the function $f : \mathbb{Q} \longrightarrow D$ defined by $f(p_i) = e_i$ is one-to-one and onto and for all i, j, $(p_i < p_j)$ if and only if $(e_i < e_j)$, therefore f is an isomorphism.   □

In our analysis of types realized in $(\mathbb{Q}, <)$ we used specific information about slopes of lines connecting points with rational coefficients. We could not argue the same way about an arbitrary countable densely ordered $(D, <)$, but now, thanks to Theorem 5.11, we know that $(\mathbb{Q}, <)$ and $(D, <)$ realize exactly the same types. Hence, Proposition 5.10 holds for $(D, <)$ as well.

In the proof of Theorem 5.11 we constructed the automorphism f by a *back-and-forth* method. In odd steps we went forward, from $\mathbb{Q}$ to D and in even steps we went back from D to $\mathbb{Q}$.

The back-and-forth method has many variants. In particular, if one only goes forward, instead of an automorphism one can construct an *embedding*. In the case of ordered structures $(D, <)$ and $(E, <)$, an embedding of $(D, <)$ into $(E, <)$ is a function $f : D \longrightarrow E$, such that for all a and b in D, if $a < b$, then $f(a) < f(b)$.

EXERCISE 5.12. Let $(D, <)$ be a countable linearly ordered set. Prove that there is an embedding of $(D, <)$ into $(\mathbb{Q}, <)$.

Let $(D, <)$ be a densely linearly ordered set. A subset E of D is *dense* in $(D, <)$, if for all a and b in D, if $a < b$, there is e in E, such that $a < e < b$. E is called *co-dense* if its complement in D is dense.

EXERCISE 5.13. Prove that there is an embedding f of $(\mathbb{Q}, <)$ into $(\mathbb{Q}, <)$ such that $f(\mathbb{Q})$ is dense and co-dense in $(\mathbb{Q}, <)$.

5.4. Types in the Field of Real Numbers

The set of real numbers is uncountable, hence $(\mathbb{R}, <)$ and $(\mathbb{Q}, <)$ are not isomorphic, but our analysis of types in $(\mathbb{Q}, <)$ can be repeated verbatim for $(\mathbb{R}, <)$ yielding the same classification of types.

Now we will turn to the field $(\mathbb{R}, +, \cdot)$. In Chapter 3, we saw that in $(\mathbb{R}, +, \cdot)$ all rational numbers are definable. Hence, each rational number has its own unique isolated type. For each rational p, let $\varphi_p(x)$ be a formula defining p. If r is irrational, then $\mathrm{tp}(r)$ includes all formulas $\forall y(\varphi_q(y) \implies y < x)$, for all q smaller than r, and $\forall y(\varphi_p(y) \implies y < p)$ for all p larger than r. Because

$\mathbb{Q}$ is dense in $\mathbb{R}$, if r and s are distinct irrational numbers it follows that $\mathrm{tp}(r) \neq \mathrm{tp}(s)$. In conclusion, any two distinct real numbers have different types; hence the identity is the only automorphism of $(\mathbb{R}, +, \cdot)$. The field $(\mathbb{R}, +, \cdot)$ is rigid.

There is more that can be said about types realized in $(\mathbb{R}, +, \cdot)$. A number r is *algebraic* if there is a non-constant polynomial $P(x)$ with integer coefficients, such that $P(r) = 0$. In other words, r is algebraic if r belongs to the set defined in $(\mathbb{R}, +, \cdot)$ by the formula $P(x) = 0$. Because solution sets of polynomial equations are finite and the ordering of the reals is definable in $(\mathbb{R}, +, \cdot)$, each algebraic number can be defined as the n-th element of the set $\{x : P(x) = 0\}$, for some $P(x)$ and n. Hence, each algebraic number has its own unique isolated type.

The question whether the types of nonalgebraic numbers are isolated is more delicate. They are not and this follows from an advanced result about the field of real numbers that implies that if r is nonalgebraic and $(\mathbb{R}, +, \cdot) \models \varphi(r)$, then the set defined by $\varphi(x)$ in $(\mathbb{R}, +, \cdot)$ contains an open neighborhood of r. Simpler examples of nonisolated types in other structures will be given in Chapter 8. The examples are simple, but to prove that they are indeed nonisolated, one needs stronger model-theoretic tools.

5.5. O-minimality of $(\mathbb{Q}, <)$

This section is not about types per se, but it is included here because the proof of the following basic result uses the techniques we discussed in this chapter.

THEOREM 5.14. $(\mathbb{Q}, <)$ *is o-minimal.*

PROOF. According to Definition 3.9, our task is to show that every set of rational numbers that is parametrically definable in $(\mathbb{Q}, <)$ is the union of finitely many open intervals and points. Let $\varphi(x, y_1, \ldots, y_n)$ be a formula, let $q_1 < \cdots < q_n$ be an increasing tuple of rational numbers, and let X be the set defined in $(\mathbb{Q}, <)$ by $\varphi(x, q_1, \ldots, q_n)$. Let I be any of the intervals: $(-\infty, q_1)$, $(q_1, q_2), \ldots, (q_{n-1}, q_n), (q_n, \infty)$. We will finish the proof if we show that if $X \cap I$ is non-empty, then $X \cap I = I$, because then X is the union of those intervals I for which $X \cap I$ is non-empty and possibly some points from the set $\{q_1, \ldots, q_n\}$.

Suppose that $I = (q_1, q_2)$. The argument for other I is similar. Suppose that p is in $X \cap I$, and let p' be any other rational number in I. Now we will proceed as in Section 5.3.2, where we classified the types in $(\mathbb{Q}, <)$. It helps to draw a picture. We can define a piecewise linear function $f : \mathbb{Q} \longrightarrow \mathbb{Q}$ made of four linear segments, such that $f(x) = x$ for all x outside I and $f(p) = p'$. This function is an automorphism of $(\mathbb{Q}, <)$ and it fixes all parameters in the definition of X. Because $(\mathbb{Q}, <) \models \varphi(p, q_1, \ldots, q_n)$, we have $(\mathbb{Q}, <) \models \varphi(p', q_1, \ldots, q_n)$; hence p' is in X. $\square$

Exactly the same argument shows that $(\mathbb{R}, <)$ is o-minimal. In fact, all densely linearly ordered sets are o-minimal and even more, the ordered field

$(\mathbb{R}, +, \cdot, <)$ is o-minimal. Those results require more preparation than we can give here. See [**Mar02**]. Later we will show that $(\mathbb{Z}, <)$ is o-minimal, so it would seem that perhaps all ordered sets are o-minimal. This is not the case, as shown by the following example.

EXAMPLE 5.15. Let $(F, <)$ be the ordered set of rational numbers of the form $n + \frac{1}{k}$ where n and k are natural numbers and $k \neq 0$, 1, with the usual ordering. The formula $\neg \exists y\, \mathrm{Succ}(x, y)$ defines the set of natural numbers in $(F, <)$; hence, $(F, <)$ is not o-minimal.

CHAPTER 6

Elementarity

In model theory, we identify isomorphic structures. This is not a strict rule and there are exceptions, especially where effective (computable) presentations of structures are considered. In any case though, isomorphism is the strongest notion of similarity. In this chapter we will examine a weaker notion— elementary equivalence.

6.1. Elementary Equivalence

Recall that the theory of a structure $\mathfrak{M}$, denoted $\mathrm{Th}(\mathfrak{M})$, is the set of all first-order sentences that are true in $\mathfrak{M}$.

DEFINITION 6.1. We say that $\mathfrak{M}$ and $\mathfrak{N}$ are *elementarily equivalent* if $\mathrm{Th}(\mathfrak{M}) = \mathrm{Th}(\mathfrak{N})$.

In other words, structures are elementarily equivalent if they are structures for the same language and you cannot tell the difference between them in a first-order way. Often, classes of structures are characterized by their particular first-order properties, and this gives many examples of structures with different theories: abelian and non-abelian groups, fields of different characteristics, finite structures of different cardinalities (this is explained in detail below), finite and infinite structures, orderings with and without a least element, dense orderings and orderings that are not dense, and many more.

By Theorem 4.6 isomorphic structures are elementarily equivalent. However if two structures are not isomorphic they may or may not be elementarily equivalent and it is not always easy to tell. An example is given in the following exercise.

EXERCISE 6.2. In Exercise 4.10, the reader was asked to show that groups $(\mathbb{Z} \times \mathbb{Z}, +)$ and $(\mathbb{Z}, +)$ are not isomorphic. Prove that they are not elementarily equivalent. It is a harder exercise. HINT: In $(\mathbb{Z}, +)$ each number is either even or odd.

For examples of structures that are elementarily equivalent but not isomorphic, we have to wait. The next theorem shows that we will not find them among structures with finite domains. First, let us note that for each natural number n, having domain of cardinality n is a first-order property. For a set X, $|X|$ is the cardinality of X.

For a finite set of formulas Φ, $\bigwedge \Phi$ is the conjunction of all formulas in Φ and $\bigvee \Phi$ is the disjunction of formulas in Φ.

PROPOSITION 6.3. *For each natural number n, there is a first-order sentence φ_n such that for each structure $\mathfrak{M}$, $\mathfrak{M} \models \varphi_n$ if and only if $|M| = n$.*

PROOF. Because we allow structures with empty domains, we need φ_0. It can be any sentence that is always false, for example $\forall x(x \neq x)$. For $n > 0$, let φ_n be be the sentence that says "there are n different elements and every element is one of them." Formally,

$$\exists x_1, \ldots, \exists x_n [\bigwedge \{x_i \neq x_j : i, j \leq n \text{ and } i \neq j\} \wedge \forall y \bigvee \{y = x_i : 1 \leq i \leq n\}].$$

$\square$

THEOREM 6.4. *Let $\mathfrak{M}$ be a structure with a finite domain. If $\mathfrak{M}$ and $\mathfrak{N}$ are elementarily equivalent, then $\mathfrak{M}$ is isomorphic to $\mathfrak{N}$.*

PROOF. Suppose, to the contrary, that $\mathfrak{M}$ and $\mathfrak{N}$ are elementarily equivalent, but not isomorphic. Let $n = |M|$, and let φ_n be the sentence defined in Proposition 6.3. Then $\mathfrak{M} \models \varphi_n$ and, because $\mathfrak{N}$ is elementarily equivalent to $\mathfrak{M}$, $\mathfrak{N} \models \varphi_n$; hence $n = |N|$.

Because we assumed that $\mathfrak{M}$ and $\mathfrak{N}$ are not isomorphic, none of the one-to-one functions $f : M \longrightarrow N$ is an isomorphism. The number of such functions is $n!$. Let $m = n!$, and let f_1, f_2, $\ldots$, f_m be a list of those functions. Let $a_1, a_2, \ldots, a_n$ be a list of all elements of M.

Because none of the functions f_k is an isomorphism, for each positive $k \leq m$ there is a formula $\varphi_k(x_1, x_2, \ldots, x_n)$ such that

$$\mathfrak{M} \models \varphi_k(a_1, a_2, \ldots, a_n),$$

and

$$\mathfrak{N} \models \neg \varphi_k(f_k(a_1), f_k(a_2), \ldots, f_k(a_n)). \qquad (*)$$

There could be such formulas φ_k with a smaller number of free variables, but we can make sure that all $x_1, \ldots, x_n$ are there by replacing φ_k with

$$\varphi_k \wedge \bigwedge \{x_i = x_i : 0 < i \leq n\}.$$

Let $\psi(x_1, \ldots, x_n)$ be

$$\varphi_1(x_1, \ldots, x_n) \wedge \varphi_2(x_1, \ldots, x_n) \wedge \cdots \wedge \varphi_m(x_1, \ldots, x_n),$$

and let Φ be

$$\exists x_1 \ldots \exists x_n \bigwedge \{x_i \neq x_j : i, j \leq n \text{ and } i \neq j\} \wedge \psi(x_1, \ldots, x_n).$$

Because for each k, $\mathfrak{M} \models \varphi_k(a_1, \ldots, a_n)$, it follows that $\mathfrak{M} \models \Phi$, and since $\mathfrak{N}$ is elementarily equivalent to $\mathfrak{M}$, $\mathfrak{N} \models \Phi$. Then there are distinct b_1, $\ldots$, b_n in N such that

$$\mathfrak{N} \models \psi(b_1, \ldots, b_n). \qquad (**)$$

Let $f : M \longrightarrow N$ be defined by $f(a_i) = b_i$, for $i = 1, \ldots, n$. Because f is one-to-one, it is one of the functions f_k on our list. Then, by $(*)$, $\mathfrak{N} \models \neg\varphi_k(b_1, b_2, \ldots, b_n)$, but that contradicts $(**)$ and this finishes the proof. $\qquad\square$

Theorem 6.4 has an interesting corollary, but before we state it, we need another note on notation that we already used informally in the previous chapter.

Note on notation: For a in M, by $(\mathfrak{M}, a)$ we denote the structure $\mathfrak{M}$ expanded by adding a as a new constant to the structure. This means that we also expand the language of $\mathfrak{M}$ by adding a new constant symbol. In this context, if $(\mathfrak{M}, a)$ and $(\mathfrak{N}, b)$ are two such expansions, where the languages of $\mathfrak{M}$ and $\mathfrak{N}$ are the same, then we assume that a and b are interpretations of the same newly added symbol. In particular, if $f : M \longrightarrow N$ is an isomorphism, then, by definition, $f(a) = b$.

Similarly, we can expand $\mathfrak{M}$ by adding a tuple $\bar{a}$ to get $(\mathfrak{M}, \bar{a})$. For any $A \subseteq M$ we can define $(\mathfrak{M}, a)_{a \in A}$, but here we have to be more careful, because we need to know which new constant symbol in the language corresponds to which element of M. A simple solution is to designate each a in A as a constant symbol for itself. This could be problematic, but works well for structures $\mathfrak{N}$ such that $M \subseteq N$, because then $(\mathfrak{N}, a)_{a \in A}$ is still well defined.

With the notation explained, we can move to the next proposition. It follows directly from definitions.

PROPOSITION 6.5. Let $\bar{a}$ be in $M^{<\omega}$ and $\bar{b}$ be in $N^{<\omega}$. Then $\mathrm{tp}^{\mathfrak{M}}(\bar{a}) = \mathrm{tp}^{\mathfrak{N}}(\bar{b})$ if and only if $(\mathfrak{M}, \bar{a})$ is elementarily equivalent to $(\mathfrak{N}, \bar{b})$.

Proposition 6.5 is not really a proposition but rather an explanation that a certain phenomenon can be described from two different perspectives. In formal logic, the roles of variables and constants are different, but in model-theoretic practice this difference is often blurred. If a is an element of M, then formally $\mathrm{tp}^{\mathfrak{M}}(a)$ is the set of *formulas* $\varphi(x)$ such that $\mathfrak{M} \models \varphi(a)$, where x is a variable, while $\mathrm{Th}(\mathfrak{M}, a)$, is the set of *sentences* $\varphi(a)$ for which $\mathfrak{M} \models \varphi(a)$, where a is a constant symbol added to the language. If a is already a constant of $\mathfrak{M}$, then the language of $\mathfrak{M}$ already has a symbol for it and in this case $\mathrm{Th}(\mathfrak{M}, a)$ is just $\mathrm{Th}(\mathfrak{M})$.

What is the point of having those two different perspectives? Everything that can be seen from one of them can also be seen from the other, after a translation. The types perspective has an algebraic character. An n-type is a set of formulas and as such it resembles a system of equations in which the unknowns are the free variables of the formulas in the type. A realization of the type is an analog of a solution of the system. We can think of a as a solution of $\mathrm{tp}^{\mathfrak{M}}(a)$ and then we can ask if there are other solutions in M, but we can also do more. If $\mathfrak{N}$ is a structure for the same language, we can ask if $\mathrm{tp}^{\mathfrak{M}}(a)$ has a solution in $\mathfrak{N}$, i.e. is $\mathrm{tp}^{\mathfrak{M}}(a)$ realized in $\mathfrak{N}$. If it is, then for some $b \in N$, $\mathrm{tp}^{\mathfrak{M}}(a) = \mathrm{tp}^{\mathfrak{N}}(b)$ and we can turn to the other perspective and conclude that

$(\mathfrak{M}, a)$ is elementarily equivalent to $(\mathfrak{N}, b)$ because now we may want to use some model-theoretic fact about these structures.

The following corollary is an example of an argument in which the two perspectives are used.

COROLLARY 6.6. Let $\mathfrak{M}$ be a structure with a finite domain, and suppose $\bar{a}$ and $\bar{b}$ are in $M^{<\omega}$. Then $\text{tp}^{\mathfrak{M}}(\bar{a}) = \text{tp}^{\mathfrak{M}}(\bar{b})$ if and only if there is an automorphism f of $\mathfrak{M}$ such that $f(\bar{a}) = \bar{b}$.

PROOF. If $\text{tp}^{\mathfrak{M}}(\bar{a}) = \text{tp}^{\mathfrak{M}}(\bar{b})$, then, by Proposition 6.5, $(\mathfrak{M}, \bar{a})$ and $(\mathfrak{M}, \bar{b})$ are elementarily equivalent and, by Theorem 6.4, they are isomorphic. If $f : M \longrightarrow M$ is an isomorphism, then, by definition, $f(\bar{a}) = \bar{b}$ and this means that f is an automorphism f of $\mathfrak{M}$ such that $f(\bar{a}) = \bar{b}$.

If there is an automorphism f of $\mathfrak{M}$ such that $f(\bar{a}) = \bar{b}$, then $(\mathfrak{M}, \bar{a})$ and $(\mathfrak{M}, \bar{b})$ are isomorphic, hence, by Theorem 4.6 they are elementarily equivalent, so, by Proposition 6.5, $\text{tp}^{\mathfrak{M}}(\bar{a}) = \text{tp}^{\mathfrak{M}}(\bar{b})$.

$\square$

6.2. Elementary Extensions

Model theory is not much different from other branches of modern mathematics. The original motivation is to get insight into "classical" structures such as the ring of integers, the fields of real numbers and of complex numbers, and Euclidean spaces. Very soon though, it turns out that it is necessary to consider these particular structures as special cases of rings, fields and metric and topological spaces. Obviously, to study all those other structures, we first need to know where they are and how to find them. Usually, one way to get them is by constructing new ones from those structures we already have. There are well developed algebraic methods to do it. We form algebraic and topological closures; we can take various sums, products, limits, quotients and such. Model theory deals with all those operations as well, but there is one that is very specific to mathematical logic: forming elementary extensions. What they are is defined next.

DEFINITION 6.7. Let $\mathfrak{M}$ and $\mathfrak{N}$ be structures with the same language.

(1) We say that $\mathfrak{N}$ is an *extension* of $\mathfrak{M}$, written $\mathfrak{M} \subseteq \mathfrak{N}$, if $M \subseteq N$, and for every n-ary relation symbol $\mathcal{R}$ in the language, $\mathcal{R}^{\mathfrak{M}} = \mathcal{R}^{\mathfrak{N}} \cap M^n$. If $\mathfrak{M} \subseteq \mathfrak{N}$, we also say that $\mathfrak{M}$ is a *substructure* of $\mathfrak{N}$.

(2) We say that $\mathfrak{N}$ is an *elementary extension* of $\mathfrak{M}$, written $\mathfrak{M} \prec \mathfrak{N}$, if $\mathfrak{M} \subseteq \mathfrak{N}$ and for all $\bar{a}$ in $M^{<\omega}$, $\text{tp}^{\mathfrak{M}}(\bar{a}) = \text{tp}^{\mathfrak{N}}(\bar{a})$. If $\mathfrak{M} \prec \mathfrak{N}$, we also say that $\mathfrak{M}$ is an *elementary substructure* of $\mathfrak{N}$.

Each structure can be regarded as an elementary extension of itself, but in these lectures all extensions will be assumed to be *proper* extensions, i.e., extensions that add new elements to the domain.

Notice that if $\mathfrak{M} \prec \mathfrak{N}$, then $\mathrm{Th}(\mathfrak{M}) = \mathrm{Th}(\mathfrak{N})$ and this is because—as discussed in the previous chapter—the type of each element contains all sentences that are true in the structure. Hence, if $\mathfrak{M} \subseteq \mathfrak{N}$ and $\mathfrak{M}$ and $\mathfrak{N}$ are not elementarily equivalent, then $\mathfrak{N}$ is not an elementary extension of $\mathfrak{M}$. By Theorem 6.4, it follows that structures with finite domains do not have elementary extensions. Other important examples are in the following chain of extensions:

$$(\mathbb{N}, +, \cdot) \subseteq (\mathbb{Z}, +, \cdot) \subseteq (\mathbb{Q}, +, \cdot) \subseteq (\mathbb{R}, +, \cdot) \subseteq (\mathbb{C}, +, \cdot).$$

None of the extensions is elementary, because the theories of all these structures are different. Each larger structure contains solutions of some polynomial equations with integer coefficients that are not solvable in smaller structures and this can be expressed in a first-order way.

In the next chapter, we will prove that every structure with an infinite domain has an elementary extension, so elementary extensions exist in abundance. For now, one simple example of an elementary extension is given in Proposition 6.10 below.

While isomorphic structures are elementarily equivalent, the converse does not hold, as the following example shows.

EXAMPLE 6.8. Let $\mathbb{N}^+$ be the set of positive natural numbers, let $\mathfrak{M}$ be $(\mathbb{N}^+, <)$, and let $\mathfrak{N}$ be $(\mathbb{N}, <)$. $\mathfrak{N}$ is an extension of $\mathfrak{M}$, but not an elementary one. For example, 1 is the least element in $\mathfrak{M}$, but not in $\mathfrak{N}$, hence $\mathrm{tp}^{\mathfrak{M}}(1) \neq \mathrm{tp}^{\mathfrak{N}}(1)$. In fact, since every number in both structures is defined by its position with respect to the least element, for each $n > 0$, $\mathrm{tp}^{\mathfrak{M}}(n) \neq \mathrm{tp}^{\mathfrak{N}}(n)$. On the other hand, $\mathfrak{M}$ is isomorphic to $\mathfrak{N}$, hence $\mathfrak{M}$ and $\mathfrak{N}$ are elementarily equivalent by Theorem 4.6.

6.3. Tarski-Vaught Test

The Tarski-Vaught test is a tool for detecting elementarity of extensions. It is a simple test, with a simple proof, however, it does not seem to be that simple to those who see it for the first time and this in itself is an interesting phenomenon. For those who teach, it is an effective litmus test to see if the students follow the course. If they can reproduce the proof on an exam, everything is fine. In practice quite often this is not the case. So please, go over the proof carefully.

THEOREM 6.9. *If $\mathfrak{N}$ is an extension of $\mathfrak{M}$, then the extension is elementary if and only if for all $n > 0$, all $\bar{a}$ in M^n and all formulas $\varphi(x, \bar{y})$, if $\mathfrak{N} \models \exists x \varphi(x, \bar{a})$, then $\mathfrak{N} \models \varphi(b, \bar{a})$ for some b in M.*

PROOF. If $\mathfrak{M} \prec \mathfrak{N}$ and $\mathfrak{N} \models \exists x \varphi(x, \bar{a})$, then $\mathfrak{M} \models \exists x \varphi(x, \bar{a})$. Thus, we only need to prove the other direction of the equivalence. Assuming that the condition in the theorem holds, we will show by induction on the complexity of $\varphi(\bar{x})$ that for all $\varphi(\bar{x})$ and all $\bar{a}$ in $M^{<\omega}$,

$$\mathfrak{M} \models \varphi(\bar{a}) \text{ iff } \mathfrak{N} \models \varphi(\bar{a}). \tag{$*$}$$

First notice that $(*)$ holds for all atomic formulas, because $\mathfrak{N}$ is an extension of $\mathfrak{M}$ (Definition 6.7).

Now assume that $(*)$ holds for all formulas of rank k, and let $\varphi(\bar{x})$ be of rank $k+1$ (Definition 4.5). It is easy to check that $(*)$ holds if φ is either of the form $\psi \wedge \theta$ or of the form $\neg\psi$. If φ is of the form $\exists y\psi(y, \bar{x})$, then

$$\mathfrak{N} \models \varphi(\bar{a}) \Longleftrightarrow \mathfrak{N} \models \psi(b, \bar{a}) \text{ for some } b \text{ in } M$$
$$\Longleftrightarrow \mathfrak{M} \models \psi(b, \bar{a}) \text{ (by the inductive assumption)}$$
$$\Longleftrightarrow \mathfrak{M} \models \varphi(\bar{a}).$$

$\square$

As an application of the Tarski-Vaught test, we will prove the following proposition.

PROPOSITION 6.10. $(\mathbb{Q}, <) \prec (\mathbb{R}, <)$.

PROOF. Suppose that for a tuple of rational numbers $\bar{q}$, $(\mathbb{R}, <) \models \exists x\varphi(x, \bar{q})$. Then

$$(\mathbb{R}, <) \models \varphi(r, \bar{q}),$$

for some r in $\mathbb{R}$. If r is rational, we are done. If r is irrational, then, as in the proof of Theorem 5.14, we find an automorphism of $(\mathbb{R}, <)$ that fixes $\bar{q}$ and such that $f(r) = p$ for some p in $\mathbb{Q}$. Then $(\mathbb{R}, <) \models \varphi(p, \bar{q})$, showing that the assumption of the Tarski-Vaught test is satisfied. $\square$

The next result is known as the *elementary chain lemma*. It is formulated for countable chains of models, in full generality it holds for chains of arbitrary length.

LEMMA 6.11. Let $\mathfrak{M}_n$ be a sequence of structures, such that, for each n, $\mathfrak{M}_n \prec \mathfrak{M}_{n+1}$. Let the domain of $\mathfrak{M}$ be $\bigcup_{n \in \mathbb{N}} M_n$, and for each relation symbol $\mathcal{R}$ of the language, let $\mathcal{R}_\mathfrak{M} = \bigcup_{n \in \mathbb{N}} \mathcal{R}_{\mathfrak{M}_n}$. Then, each $\mathfrak{M}_n$ is an elementary substructure of $\mathfrak{M}$.

EXERCISE 6.12. Use Tarski-Vaught test to prove Lemma 6.11. If you are familiar with basic set theory, formulate and prove the result for arbitrary chains of models.

6.4. End Note on Structures with Finite Domains

Theorem 6.4 is formulated for structures with *finite domains* and not just *finite structures*. The reason is that a structure with a finite domain can be infinite in the sense that it can have infinitely many relations.

EXAMPLE 6.13. For each set X of natural numbers we will define a structure $\mathfrak{M}_X$ whose domain is $\{a\}$. For each $n > 0$, $\{a\}^n$ is the one-element set containing the n-tuple $(a, \ldots, a)$. The language of each $\mathfrak{M}_X$ is the set of relation symbols $\mathcal{R}_n$, one for each $n > 0$, and the arity of $\mathcal{R}_n$ is n. The relation symbol $\mathcal{R}_n$ is interpreted in $\mathfrak{M}_X$ as $\{a\}^n$ if n is in X and as the empty set otherwise.

EXERCISE 6.14. Show that if $X \neq Y$ then $\mathfrak{M}_X$ and $\mathfrak{M}_Y$ are not isomorphic.

$\{\mathfrak{M}_X : X \subseteq \mathbb{N}\}$ is a set of cardinality $2^{\aleph_0}$ of non-isomorphic models with a one-element domain. Theorem 6.4 tells us that any two of those models are not elementarily equivalent, but this is easy to see directly.

EXERCISE 6.15. Without appealing to Theorem 6.4, show that if $X \neq Y$ then $\mathrm{Th}(\mathfrak{M}_X) \neq \mathrm{Th}(\mathfrak{M}_Y)$.

CHAPTER 7

Compactness

The compactness theorem is a cornerstone of model theory. It says that if every finite subset of a theory has a model, then the theory has a model. The theorem follows from the famous completeness theorem proved by Kurt Gödel in 1929. The completeness theorem says that every consistent theory has a model. There is something suspicious here, and that is because two chapters ago we defined consistent theories as those that have models. The explanation is that there are two notions of consistency based on two notions of logical inference: a semantic and a syntactic one.

After a short discussion of the two forms of inference, a proof of the compactness theorem is given. The purpose here is not just to validate the result. The main ingredient of the proof is a construction, due to Leon Henkin, that has many other applications. All steps of the construction are explained in detail. We follow with some applications of the theorem and a short optional section on the compactness of topological spaces of complete theories.

7.1. Semantic and Syntactic Inference

We will define what it means that one first-order sentence *implies* another. The definition does not involve any notion of proof. Instead, it is grounded in set theory and Tarski's definition of truth, as it refers to *all* structures that are models of a particular sentence.

DEFINITION 7.1. For first-order sentences φ and ψ, we say that φ *implies* ψ if for every $\mathfrak{M}$, if $\mathfrak{M} \models \varphi$, then $\mathfrak{M} \models \psi$. A sentence φ is a *consequence* of a theory T, or T *proves* φ, if for every $\mathfrak{M}$, if $\mathfrak{M} \models T$, then $\mathfrak{M} \models \varphi$.

We will write $\varphi \models \psi$, if φ implies ψ and $T \models \psi$, if ψ is a consequence of T.

Let us see how semantic inference works in a simple, but important case. In mathematics, any inconsistent theory proves all sentences. This is well-known, but not very intuitive. Let us see how it is justified in terms of semantic inference. Suppose T is inconsistent. According to our definition, this means that T has no model. Then, for any sentence φ, the statement "if $\mathfrak{M}$ is a model of T then $\mathfrak{M}$ is a model of φ" is vacuously true, because there are no such models. Hence, $T \models \varphi$. This argument is based on the truth value assignment to propositional formulas of the form $p \implies q$. In classical logic, the conditional $p \implies q$ is defined as $\neg p \lor q$. It follows that its truth value is "true" whenever p is false.

Syntactic inference refers to a formal derivation in a proof system based on axioms of first-order logic and specific rules of proof. If $\mathcal{P}$ is such a system and T is a theory, then $T \vdash_{\mathcal{P}} \varphi$ means that there is a proof of φ that uses sentences in T as axioms. Gödel proved his completeness theorem for the Hilbert Ackerman proof system (see [**Ken18**] for historical details), but it can be proved for any proof system that satisfies some natural conditions. In this sense, it is not one theorem, but many—one for each proof system $\mathcal{P}$.

One of those natural conditions is *soundness*. A proof system is *sound* if for all theories T and all first-order sentences φ if $T \vdash_{\mathcal{P}} \varphi$, then $T \models \varphi$. In other words, if there is a formal proof of φ from T, then φ is true in all models of T.

THEOREM 7.2 (Completeness Theorem). *Let T be a theory. Then for every sentence φ in the language of T, $T \models \varphi$ if and only if $T \vdash_{\mathcal{P}} \varphi$.*

For a proof and a discussion of deduction and completeness, see [**Doe96**, Appendix A].

Finally let us compare the semantic and syntactic version of the following theorem.

THEOREM 7.3. *Every consistent theory has an extension to a complete consistent theory.*

In the semantic version, it is not really a theorem: If T is a consistent theory, then, by definition, it has a model $\mathfrak{M}$, then $T \subseteq \mathrm{Th}(\mathfrak{M})$ and $\mathrm{Th}(\mathfrak{M})$ is consistent and complete. The syntactic version follows similarly, after one first evokes the completeness theorem. A direct proof-theoretic argument can also be given, based on the fact that if a theory T is consistent, then for every first-order sentence φ, either $T \cup \{\varphi\}$ or $T \cup \{\neg\varphi\}$ is consistent.

7.2. The Compactness Theorem

There are many proofs of the compactness theorem. From the assumption that every finite set of sentences of a given first-order theory has a model, we will derive the conclusion that the whole theory has a model. Because every finite fragment of T has a model, then one could try to somehow assemble all those models together to build a model of T. There is a way to do it, but we will follow a different route. In 1947 Leon Henkin gave a simplified proof of the completeness theorem. We will use Henkin's method, to build a model of T directly. Because what we start with is just a set of syntactic objects—sentences of T—one may wonder what will this model of T be made of. It will be made of what we have, i.e., some elements of syntax, namely constant symbols. I heard this insight from professor Andrzej Mostowski when I was a student in Warsaw. I thought it was quite profound.

We will say that a theory T is *finitely consistent*, written $\mathrm{Confin}(T)$, if every finite subset of T has a model. Now the compactness theorem can be formulated as follows.

THEOREM 7.4 (Compactness Theorem). *Every finitely consistent theory is consistent.*

We will only prove the compactness theorem for countable languages. This version suffices for most applications that we will discuss here. The proof of the theorem in the general case is similar, but it uses more advanced set theory.

In preparation for the proof, we need two general lemmas about finite consistency. If φ is a sentence, to simplify notation, we will write $T + \varphi$ instead of $T \cup \{\varphi\}$.

LEMMA 7.5. If $\mathrm{Confin}(T)$, then for every sentence φ either $\mathrm{Confin}(T + \varphi)$ or $\mathrm{Confin}(T + \neg\varphi)$.

PROOF. Suppose that $T + \varphi$ is not finitely consistent. Then there is a finite $T' \subseteq T$ such that $T' + \varphi$ has no model. Let $T'' \subseteq T$ be finite. $T' \cup T''$ has a model $\mathfrak{M}$. Because $T' + \varphi$ has no model, we have $\mathfrak{M} \models \neg\varphi$. So $\mathfrak{M}$ is a model of $T'' + \neg\varphi$. This shows that $T + \neg\varphi$ is finitely consistent. □

LEMMA 7.6. If $\mathrm{Confin}(T + \exists x\varphi(x))$ and c is a constant symbol that does not occur either in T or in $\varphi(x)$, then $\mathrm{Confin}(T + \varphi(c))$.

PROOF. Let $T' \subseteq T$ be finite. $T' + \exists x\varphi(x)$ has a model $\mathfrak{M}$. Because $\mathfrak{M} \models \exists x\varphi(x)$, there is an a in M such that $\mathfrak{M} \models \varphi(a)$. Because neither T' nor $\varphi(x)$ mentions c, we can interpret c in $\mathfrak{M}$ as a. Then $\mathfrak{M} \models \varphi(c)$, proving $\mathrm{Confin}(T + \varphi(c))$. □

The above proof looks quite simple, but there is something swept under the rug. To complete the proof, please do the following exercise.

EXERCISE 7.7. Use induction on the complexity of $\varphi(x)$ to verify the claim that $\mathfrak{M} \models \varphi(c)$ in the last line of the proof above.

PROOF. The proof of the compactness theorem begins now. Let T be a finitely consistent theory in a countable language $\mathcal{L}_T$.

We extend $\mathcal{L}_T$ to $\mathcal{L}_T(C)$ by adding a countable set of new constant symbols $C = \{c_n : n \in \mathbb{N}\}$. Because $\mathcal{L}_T(C)$ is countable, the set of all formulas of $\mathcal{L}_T(C)$ is still countable as it is generated from the atomic formulas by a computable process, i.e., all formulas of $\mathcal{L}_T(C)$ can be obtained by systematic step-by-step process of building Boolean combinations and affixing quantifiers. Let $\{\varphi_n : n \in \mathbb{N}\}$ be an enumeration of all of sentences of $\mathcal{L}_T(C)$. Not any enumeration will do. A small technical point is that we have to make it so that the new constant symbols that can occur in φ_n are among $c_0, \ldots, c_{n-1}$. In particular, there are no new constant symbols in φ_0. There are many ways to do it. Here is one. Let $\{\psi_n : n \in \mathbb{N}\}$ be any enumeration of the set of sentences of $\mathcal{L}_T(C)$. Let φ_0 be the sentence ψ_i with the least index i that has no new constants. Then proceed inductively. If φ_k are defined for $k \leq n$, let φ_{n+1} be the sentence ψ_i with the least index i in the set $\{\psi_n : n \in \mathbb{N}\} \setminus \{\varphi_0, \ldots, \varphi_n\}$ that only has constants among $c_0, \ldots, c_{n-1}$. Convince yourself that $\{\varphi_n : n \in \mathbb{N}\}$ thus defined satisfies the requirement.

We will extend T to a complete theory H in $\mathcal{L}_T(C)$. We will proceed inductively. At each step we will have a finite set T_n of sentences satisfying the inductive assumption $\mathrm{Confin}(T \cup T_n)$. For $n = 0$, we start with just T, so T_0 is empty.

Suppose that T_n has been defined. To define T_{n+1} we consider φ_n. There are three cases:

(1) $\mathrm{Confin}(T \cup T_n + \varphi_n)$ and φ_n is not of the form $\exists x \psi(x)$. Then $T_{n+1} = T_n + \varphi_n$.
(2) $\mathrm{Confin}(T \cup T_n + \varphi_n)$ and φ_n is of the form $\exists x \psi(x)$. Then $T_{n+1} = T_n + \varphi_n + \psi(c_n)$.
(3) $\neg\, \mathrm{Confin}(T \cup T_n + \varphi_n)$. Then $T_{n+1} = T_n + \neg\varphi_n$.

By Lemmas 7.5 and 7.6, we have $\mathrm{Confin}(T \cup T_{n+1})$.

Let $H = T \cup \bigcup_{n \in \mathbb{N}} T_n$. In the construction, we have guaranteed that H is complete and finitely consistent.

We do not have our model of T yet, but we are close. We will use C to build a domain. Its elements will be equivalence classes of an equivalence relation on C. We will say that c and d is C are equivalent, written $c \sim d$, if the sentence $c = d$ is in H. Let us check that it indeed is an equivalence relation. It is a "routine argument," but it has not been routine in this book yet, so let us take a look at details.

To check reflexivity, let c be a constant symbol. Because H is a complete set of sentences, for each c, either $c = c$ or $\neg(c = c)$ must be in H. It can't be $\neg(c = c)$, because then $\{\neg(c = c)\}$ is a finite subset of H, so it would have to have a model, but it has not. So $c = c$ is in H; hence $c \sim c$. For symmetry, almost the same argument shows that if $c \sim d$, then $d \sim c$. For transitivity, let us assume that $c \sim d$ and $d \sim e$. So $c = d$ and $d = e$ are in H. Because $\{c = d, d = e, \neg(c = e)\}$ has no model, by completeness, $c = e$ must be in H, so $c \sim e$.

For c in C, let $[c]$ be the equivalence class of c, i.e., $[c] = \{d : c \sim d\}$. The domain of our model $\mathfrak{M}$ will be $\{[c] : c \in C\}$. For any n-ary relation symbol $\mathcal{R}$ in $\mathcal{L}_T$, we define

$$([d_1], \ldots, [d_n]) \in \mathcal{R}_{\mathfrak{M}} \text{ iff } R(d_1, \ldots, d_n) \in H. \tag{$*$}$$

If $\mathcal{L}_T$ has its own constant symbols, they also must be interpreted in $\mathfrak{M}$. That has already been taken taken care of by H as follows. If e is a constant symbol in $\mathcal{L}_T$, then one of the sentences φ_n is $\exists x(e = x)$. Because $\mathrm{Confin}(T \cup T_n + \varphi_n)$, $T_{n+1} = T_n + \varphi_n + (e = c_n)$. Hence, the only choice for $e_{\mathfrak{M}}$ is $[c_n]$.

We come to the tedious part of the proof, we need to check that the definition $(*)$ above is correct. On the left hand side we have a statement about equivalence classes, on the right hand side a statement about particular constants. We need to check that if for $i = 1, \ldots, n$, $d_i \sim e_i$, then

$$R(d_1, \ldots, d_n) \in H \Longleftrightarrow R(e_1, \ldots, e_n) \in H.$$

Suppose that all the sentences $d_i = e_i$, and $R(d_1, \ldots, d_n)$ are in H, but $R(e_1, \ldots, e_n)$ is not. Then, because H is complete, $\neg R(e_1, \ldots, e_n)$ must be in H. Then

$$\{d_i = e_i : i = 1, \ldots, n\} \cup \{R(d_1, \ldots, d_n), \neg R(e_1, \ldots, e_n)\}$$

is a finite subset of H, so it should have a model, but clearly it does not; hence $R(e_1, \ldots, e_n)$ must be in H.

Now comes the (long) punch line. We will show that if the equivalence $(*)$ holds for atomic formulas, then it also holds for all formulas of $\mathcal{L}_T$ (without the new constants), i.e., for all formulas $\varphi(\bar{x})$ of T and all $d_1, \ldots, d_n$ in C,

$$\mathfrak{M} \models \varphi([d_1], \ldots, [d_n]) \text{ iff } \varphi(d_1, \ldots, d_n) \in H. \qquad (**)$$

As usual, the proof is by induction on the rank of $\varphi(\bar{x})$. The atomic case of $(**)$ is $(*)$ above.

Suppose that $\varphi(\bar{x})$ is $\psi(\bar{x}) \wedge \theta(\bar{x})$ and $(**)$ holds for $\psi(\bar{x})$ and $\theta(\bar{x})$. Then for all $[d_1], \ldots, [d_n]$,

$$
\begin{aligned}
\mathfrak{M} \models \varphi([d_1], \ldots, [d_n]) &\Longleftrightarrow \mathfrak{M} \models \psi([d_1], \ldots, [d_n]) \wedge \theta([d_1], \ldots, [d_n]) \\
&\Longleftrightarrow \mathfrak{M} \models \psi([d_1], \ldots, [d_n]) \text{ and } \mathfrak{M} \models \theta([d_1], \ldots, [d_n]) \\
&\Longleftrightarrow \psi(d_1, \ldots, d_n) \in H \text{ and } \theta(d_1, \ldots, d_n) \in H \text{ by } (**) \\
&\Longleftrightarrow \varphi(d_1, \ldots, d_n) \in H \text{ by completeness and finite}
\end{aligned}
$$
consistency of H .

The case for negation is similar. Let us take a look at the existential quantifier case. Suppose that $\varphi(\bar{x}) = \exists y \psi(y, \bar{x})$ and and $(**)$ holds for $\psi(y, \bar{x})$. Then, for all $[d_1], \ldots, [d_n]$,

$$
\begin{aligned}
\mathfrak{M} \models \varphi([d_1], \ldots, [d_n]) &\Longleftrightarrow \mathfrak{M} \models \exists y \psi(y, [d_1], \ldots, [d_n]) \\
&\Longleftrightarrow \mathfrak{M} \models \psi([c], [d_1], \ldots, [d_n]) \text{ for some } c \text{ in } C \\
&\Longleftrightarrow \psi(c, d_1, \ldots, d_n) \in H \text{ (by } (**)) \\
&\Longrightarrow \exists y \psi(y, d_1, \ldots, d_n) \in H \text{ by completeness and finite}
\end{aligned}
$$
consistency of H
$$\Longleftrightarrow \varphi(d_1, \ldots, d_n) \in H.$$

The next to the last arrow in the derivation above points only in one direction. To reverse it, we use the crucial feature of the construction. If $\exists x \psi(x, d_1, \ldots, d_n)$ is in H, it must have been added at some stage to one of the T_n's, but then $\psi(c_n, d_1, \ldots, d_n)$ was added to T_n as well. This is all we need to run the argument in the other direction, form the bottom to the top.

So why is $\mathfrak{M}$ a model of T? That is because T is a subset of H and, by $(**)$ above, $\varphi \in H$ if and only if $\mathfrak{M} \models \varphi$. $\qquad \square$

The proof that we just gave has an interesting corollary.

COROLLARY 7.8. If a theory T in a countable language has a model, then it has a model that is either countable or finite.

PROOF. If T has a model, then this model is a model of each finite subset of T. We can apply the proof of Theorem 7.4 to obtain a model of T whose domain is built of equivalence classes of the countable set of constants; hence it is either finite or countable.   □

Corollary 7.8 shows that if Zermelo-Fraenkel axiomatic theory ZF has a model, then it has a countable model. This result is known as the Löwenheim-Skolem paradox or Skolem paradox, as it is usually derived from the Löwenheim-Skolem theorem (Theorem 11.22).

7.2.1. Existence of elementary extensions. In the next chapter we will explore consequences of another simple corollary of the compactness theorem. For the proof of the corollary we will need the following observation. For a structure $\mathfrak{M}$, $(\mathfrak{M}, a)_{a \in M}$ is the expansion of $\mathfrak{M}$ obtained by adding all elements of the domain of $\mathfrak{M}$ as constants. Let $T = \mathrm{Th}((\mathfrak{M}, a)_{a \in M})$. It follows directly from definitions that if $\mathfrak{N} \models T$ then $\mathfrak{M} \prec \mathfrak{N}$. Here we take advantage of the fact that in model theory we can identify isomorphic structures. Let us take a close look at what is going on here.

When we formed the expansion $(\mathfrak{M}, a)_{a \in M}$, we added to the language constant symbols for all elements a of M. For simplicity of notation, we can use a both as a formal symbol and an informal name of an element of M (see the note on notation after the proof of Theorem 6.4). In other words, for each a in M, $a_{\mathfrak{M}} = a$, where a on the left is used as a constant symbol. Now, if $\mathfrak{N} \models T$, then each symbol a has an interpretation $a_{\mathfrak{N}}$ in N. Let $\mathfrak{M}'$ be the structure whose domain M' is $\{a_{\mathfrak{N}} : a \in M\}$ and whose relations are restrictions to M' of the relations of $\mathfrak{N}$. It follows from definitions that: (1) the function $f : M \longrightarrow M'$ defined by $f(a) = a_{\mathfrak{N}}$ is an isomorphism between $\mathfrak{M}$ and $\mathfrak{M}'$; (2) $\mathfrak{M}' \prec \mathfrak{N}$. So, $\mathfrak{M}$ is isomorphic to an elementary submodel of $\mathfrak{N}$. In this sense, we can say that $\mathfrak{M} \prec \mathfrak{N}$.

To be even more precise, one can use (1) and (2) above to define an elementary extension $\mathfrak{K}$ of $\mathfrak{M}$ by extending M by the set $B = N \setminus M'$ so that the domain of $\mathfrak{K}$ becomes $M \cup B$ and for each relation symbol $\mathcal{R}$ in the language, defining $\mathcal{R}_{\mathfrak{K}}$ by: for all $a_1, \dots, a_m$ in M and all $b_1, \dots, b_n$ in B, where $m + k$ is the arity of $\mathcal{R}$,

$$(a_1, \dots, a_m, b_1, \dots, b_n) \in \mathcal{R}_{\mathfrak{K}} \text{ iff } (f(a_1), \dots, f(a_m), b_1, \dots, b_n) \in \mathcal{R}_{\mathfrak{N}}. \quad (*)$$

Actually, the definition of $\mathcal{R}_{\mathfrak{K}}$ is not precise enough yet. In $(*)$ the tuples $\bar{a}$ from M and $\bar{b}$ from B are separated in a special way, so this is just an example of how $\mathcal{R}_{\mathfrak{K}}$ is defined for some formulas. A complete definition must cover all possible arrangements of $\bar{a}$ and $\bar{b}$. Notationally it becomes quite cumbersome, so we will not do it.

EXERCISE 7.9. Verify claims (1) and (2) in the paragraph above.

Now comes the promised corollary.

COROLLARY 7.10. Every structure with an infinite domain has a proper elementary extension.

PROOF. For a given $\mathfrak{M}$, let c be a constant symbol that is not in the language of $\mathfrak{M}$. Let T be the theory

$$\text{Th}((\mathfrak{M}, a)_{a \in M}) \cup \{c \neq a : a \in M\}.$$

We will show that T is finitely consistent. Let $T' \subseteq T$ be finite, and let $\{a_1, \ldots, a_n\}$ the set of all constant symbols in T' other that c, and let b in M be any element that is not in $\{a_1, \ldots, a_n\}$. Let $\mathfrak{M}'$ be $(\mathfrak{M}, a_1, \ldots, a_n, b)$, with $c_{\mathfrak{M}'} = b$. Then $\mathfrak{M}' \models T'$, because along with some sentences that are true in M, T' only requires that the interpretation of c is not among $a_1, \ldots, a_n$.

By the compactness theorem, T has a model $\mathfrak{N}$. Because $\text{Th}((\mathfrak{M}, a)_{a \in M}) \subseteq T$, by the remarks preceding the corollary, $\mathfrak{N}$ is an elementary extension of $\mathfrak{M}$. $\qquad\square$

Because Corollary 7.10 is formulated for all models, for the proof we need the compactness theorem for languages of arbitrarily large cardinalities.

7.2.2. Finiteness theorem. In formal systems, proofs are either sequences or more complex combinatorial objects. They are all by the nature of the system, finite. For such systems, it is an obvious observation that if $T \vdash_{\mathcal{P}} \varphi$, then there is a finite $T' \subseteq T$ such that $T \vdash_{\mathcal{P}} \varphi$. The next theorem is a semantic analog of this observation and it is nontrivial. Essentially, it is the compactness theorem in disguise.

THEOREM 7.11 (Finiteness Theorem). *If $T \models \varphi$, then there is a finite $S \subseteq T$ such that $S \models \varphi$.*

PROOF. First observe that, by the compactness theorem, if T is inconsistent, then some finite fragment S is inconsistent, i.e., it has no model. Then, for every φ, $S \models \varphi$.

Suppose now that T is consistent, $T \models \varphi$, but there is no finite $S \subseteq T$ such that $S \models \varphi$. Then any such S has a model $\mathfrak{M}$ such that $\mathfrak{M} \models \neg\varphi$. By the compactness theorem, $T + \neg\varphi$ has a model and this is a contradiction. $\qquad\square$

EXERCISE 7.12. Derive the compactness theorem from the finiteness theorem.

7.3. What is Compact Here?

This section is for the readers who are familiar with basic topology.

The finiteness theorem has a flavor of topological compactness. This remark can be made precise. A topological space is *compact* if each open cover of the space has a finite subcover or, equivalently, every collection of closed sets with the finite intersection property has a nonempty intersection.

To every first-order language $\mathcal{L}$, we can associate a topological space $\tau_{\mathcal{L}}$ in which points are complete and consistent $\mathcal{L}$-theories. The basic open sets of $\tau_{\mathcal{L}}$ are the sets of the form

$$[\varphi] = \{T \in \tau_{\mathcal{L}} : \varphi \in T\},$$

where φ is an $\mathcal{L}$-sentence. For example, $[\varphi \vee \neg\varphi] = \tau_{\mathcal{L}}$ and $[\varphi \wedge \neg\varphi] = \varnothing$.

To verify that $\tau_{\mathcal{L}}$ is a well-defined topological space we need to show that the intersection of two basic open sets is basic open. For given φ and ψ in $\mathcal{L}$ let $B = [\varphi] \cap [\psi]$. Then,

$$B = \{T \in \tau_{\mathcal{L}} : \varphi \in T \text{ and } \psi \in T\}.$$

Because T is consistent and complete, both φ and ψ are in T if and only if $\varphi \wedge \psi$ is in T; hence $B = [\varphi \wedge \psi]$, showing that it is basic open.

Next we will show that $\tau_{\mathcal{L}}$ is a Hausdorff space, i.e., for all T and S in $\tau_{\mathcal{L}}$, if $T \neq S$ then there are disjoint open sets U and V such that $T \in U$, $S \in V$. Indeed, if $T \neq S$, then there is a sentence φ such that $\varphi \in T$ and $\neg\varphi \in S$. So $T \in [\varphi]$, $S \in [\neg\varphi]$ and $[\varphi] \cap [\neg\varphi] = \varnothing$.

It is also easy to see that in $\tau_{\mathcal{L}}$ each basic open set is closed. Indeed, for each φ, the complement of $[\varphi]$ is the set of all theories that do not include φ, and, because the theories are complete, it is the basic open set $[\neg\varphi]$.

So now we are ready for the main theorem.

THEOREM 7.13. $\tau_{\mathcal{L}}$ is compact.

PROOF. We need to show that every open cover of $\tau_{\mathcal{L}}$ has a finite subcover. Let X be a cover of $\tau_{\mathcal{L}}$. Because every open set is the union of a collection of basic open sets, let us assume that all sets in X are basic open, i.e., $X = \{[\varphi_i] : i \in I\}$, for some index set I. To get a contradiction, suppose that X has no finite subcover.

Let $T = \{\neg\varphi_i : i \in I\}$. We will show that T is finitely consistent. Let $S = \{\neg\varphi_{i_1}, \ldots, \neg\varphi_{i_n}\}$ be a subset of T. Let

$$A = [\varphi_{i_1}] \cup \cdots \cup [\varphi_{i_n}] = [\varphi_{i_1} \vee \cdots \vee \varphi_{i_n}],$$

and let $\bar{A}$ be the complement of A in $\tau_{\mathcal{L}}$. Because X has no finite cover, $A \neq \tau_{\mathcal{L}}$. Hence,

$$\bar{A} = [\neg\varphi_{i_1} \wedge \cdots \wedge \neg\varphi_{i_n}] \neq \varnothing.$$

The last line implies that there is a consistent and complete theory extending S, proving the claim that T is finitely consistent. By the compactness theorem, T is consistent, so T is an element of $\tau_{\mathcal{L}}$ that does not belong to $[\varphi_i]$ for any i in I, contradicting that X is a cover. $\square$

EXERCISE 7.14. Derive the compactness theorem from Theorem 7.13.

CHAPTER 8

Nonstandardness

In model theory, the term "nonstandard" does not have a well-defined general meaning, but there are nonstandard models of arithmetic, and nonstandard models of set theory. We will discuss these models in a moment, but let us start with some general preliminaries.

Recall that an element a is definable in a structure $\mathfrak{M}$ if $\{a\}$ is definable in $\mathfrak{M}$. This means that there is a formula $\varphi(x)$ of the language of $\mathfrak{M}$ such that

$$\mathfrak{M} \models \varphi(a) \wedge \forall x[\varphi(x) \Longrightarrow x = a].$$

It follows that if a is definable in $\mathfrak{M}$ then it is also definable in any elementary extension of $\mathfrak{M}$. Moreover, if $\mathfrak{M} \prec \mathfrak{N}$, and $\varphi(x)$ defines a in $\mathfrak{N}$, then $\mathfrak{N} \models \exists x \varphi(x)$, by elementarity, the same sentence holds in $\mathfrak{M}$ and it follows that a is in M, because otherwise there would be at least two elements of N satisfying $\varphi(x)$. This proves the following proposition.

PROPOSITION 8.1. If $\mathfrak{M} \prec \mathfrak{N}$, then the set of definable elements of $\mathfrak{N}$ is the same as the set of definable elements of $\mathfrak{M}$.

The next proposition follows directly from Theorem 4.6.

PROPOSITION 8.2. If $f : M \longrightarrow N$ is an isomorphism, then, for all a in M, a is definable in $\mathfrak{M}$ if and only if $f(a)$ is definable in $\mathfrak{N}$.

For the rest of this chapter, we will need a bit of set theory.

DEFINITION 8.3. A linearly ordered set $(A, <)$ is *well-ordered* if every nonempty subset of A has a least element.

The natural numbers are well-ordered, but the integers are not. In set theory, using the axiom of choice one can prove the well-ordering principle: every set can be well-ordered.

8.1. Nonstandardness in Arithmetic

The *standard model of arithmetic* is the set of natural numbers with addition and multiplication, $(\mathbb{N}, +, \cdot)$. The term *nonstandard model of arithmetic* is used for any structure for the same language that is not isomorphic to the standard model, but resembles it. To resemble the standard model may mean different things. A nonstandard model $\mathfrak{M}$ can be elementarily equivalent to $(\mathbb{N}, +, \cdot)$, or can be a model of the axioms of Peano Arithmetic PA, or a version of those axioms that can be weaker or stronger than PA. The closest $\mathfrak{M}$ can get to the standard model is by being its elementary extension. An elementary extension shares with the standard model all properties that are first-order expressible in its language. Let us examine one example.

Let $\mathfrak{N}$ be the standard model, and let $P(x)$ be the formula

$$x > 1 \land \forall y, z(x = y \cdot z \implies (x = y \lor x = z)).$$

For each natural number n, $\mathfrak{N} \models P(n)$ if and only if n is prime; hence

$$\mathfrak{N} \models \forall x \exists y [x < y \land P(y)]. \tag{1}$$

Let $\mathfrak{M} = (M, +, \cdot)$ be an elementary extension of $\mathfrak{N}$. Then (1) holds in $\mathfrak{M}$ and this means that the set defined by $P(x)$ in $\mathfrak{M}$ is unbounded in M. We will call the elements of $M \setminus \mathbb{N}$ *nonstandard primes*.

Suppose there is a nonstandard prime c, such that $c + 2$ is also prime. Then, for each standard natural number n,

$$\mathfrak{M} \models \exists x [n < x \land P(x) \land \forall y (y = x + 2 \implies P(y))]. \tag{2}$$

By elementarity, (2) holds in $\mathfrak{N}$, therefore there is a standard p such that $n < p$ and both p and $p + 2$ are prime. Since n was arbitrary, we have shown that there are infinitely many twin primes.

Despite recent progress, the twin primes conjecture remains open. If the conjecture is true, then there are unboundedly many nonstandard twin primes in every elementary extension of $\mathfrak{N}$, if the conjecture is false and

$$T = \{(3, 5), (5, 7), (11, 13), \ldots, (p, p + 2)\}$$

is the set of all (standard) pairs of twin primes, then T is the set of all twin primes in every elementary extension of $\mathfrak{N}$.

The ordering of the natural numbers is definable in the standard model and—as we have seen in Chapter 5.2—every natural number is definable in $(\mathbb{N}, <)$. If $\mathfrak{M}$ is an elementary extension of the standard model, by Proposition 8.1, all elements of $M \setminus \mathbb{N}$ are undefinable; hence, by Proposition 8.2, $\mathfrak{M}$ is not isomorphic to $(\mathbb{N}, +, \cdot)$, so it is nonstandard.

Models of arithmetic will be discussed in more detail in Chapter 12. In this chapter we will take a look at the reducts of the standard model $(\mathbb{N}, <)$ and $(\mathbb{N}, +)$ and we will prove some useful results about their elementary extensions.

8.2. Nonstandardness in Set Theory

This section requires familiarity with the basics of axiomatic set theories.

There is no standard model of set theory. To begin with, there are many axiomatic systems to choose from. In mathematical practice, it is commonly accepted that most of known mathematics can be formalized in ZFC, the Zermelo-Fraenkel axioms with the axiom of choice, so one could select ZFC as standard, but what about its models? It may be disturbing that we do not even know if there are any. Why, you would ask, doesn't every syntactically consistent theory have a model? Yes, this is what the completeness theorem says, but how do we know that there is no formal derivation of inconsistency from the axioms of ZFC? While the completeness theorem can be formulated and proved as a theorem of ZFC, Gödel's second incompleteness theorem implies that we cannot prove in ZFC that ZFC has a model. One can prove that models of ZFC exists assuming stronger axioms, but then the problem persists; how do we know that the stronger axioms are consistent with ZFC? Moreover, research in axiomatic set theory has shown that if ZFC has a model, then it has a whole spectrum of nonisomorphic models and there is little indication of which one could be considered standard. Still, there is a notion of nonstandardness.

From now on, let us assume that ZFC is consistent, as is commonly done in mathematics. The first-order language of set theory has only one binary relation symbol $\in$. Let $\mathfrak{V} = (V, \in^{\mathfrak{V}})$ be a model of ZFC. There is a formula $\varphi_{\mathbb{N}}(x)$ that defines the set of natural numbers in each model of ZFC. Let ν be the unique element in V, such that $\mathfrak{V} \models \varphi_{\mathbb{N}}(\nu)$. Then

$$\mathbb{N}^{\mathfrak{V}} = \{n : n \in V \text{ and } \mathfrak{V} \models n \in \nu\}$$

is the set of natural numbers of $\mathfrak{V}$.

The set of standard natural numbers is well-ordered and the fact that the set defined by $\varphi_{\mathbb{N}}(x)$ is well-ordered by the usual ordering of natural numbers is a theorem of ZFC. This means that $\mathfrak{V}$ thinks that $(\nu, <)$ is well-ordered. But looking from outside of the model, this $\mathbb{N}^{\mathfrak{V}}$ may or may not be well-ordered. In any case, we can use $\mathfrak{V}$ to prove that there is a model $\mathfrak{W} = (W, \in^{\mathfrak{W}})$ of ZFC such that $\mathbb{N}^{\mathfrak{W}}$, i.e., the set of natural numbers of $\mathfrak{W}$, is not well-ordered. Of course, if $\mathbb{N}^{\mathfrak{V}}$ is not well-ordered, there is nothing to prove.

Consider the following theory T in the language of ZFC expanded by the set $\{c_n : n \in \mathbb{N}\}$ of constant symbols.

$$\mathsf{ZFC} \cup \{\forall x[\varphi_{\mathbb{N}}(x) \implies c_n \in x \wedge c_{n+1} < c_n] : n \in \mathbb{N}\}.$$

Every finite fragment of T can be realized in $\mathfrak{V}$ by interpreting the constants in it by a suitably long finite sequence of elements in $\mathbb{N}^{\mathfrak{V}}$. This shows that T is finitely consistent and, by the compactness theorem, it has a model $\mathfrak{W}$. The natural numbers of $\mathfrak{W}$ contain an infinite descending sequence, hence they are not well-ordered.

Models of ZFC whose sets of natural numbers are not well-ordered, such as $\mathfrak{W}$ above, are called ω-*nonstandard*. In set theory, everything is a set, in particular, every structure can be represented as a single set by a suitable coding. Moreover, there is a first-order formula $\varphi_{\mathrm{WO}}(x)$ expressing that x is (or codes) a well-ordered set. In general, a model $\mathfrak{V}$ of ZFC is nonstandard if there is an a in V such that $\mathfrak{V} \models \varphi_{\mathrm{WO}}(a)$, but the set $\{x : \mathfrak{V} \models x \in a\}$ is not well-ordered. It can be shown that every nonstandard model of ZFC has an ordinal number that is not well ordered

In the "real world," for every ordinal number α, the set of ordinal numbers smaller than α is well-ordered. There is a first-order formula $\varphi_{\mathrm{ORD}}(x)$, that defines the set of ordinal numbers in every model of ZFC. For every model $\mathfrak{V}$, $\mathfrak{V} \models \forall \varphi_{\mathrm{ORD}}(x) \implies \varphi_{\mathrm{WO}}(x)$, regardless whether $\mathfrak{V}$ is standard or not.

However, if $\mathfrak{V}$ is nonstandard, then there is an α such $\mathfrak{V} \models \varphi_{\mathrm{WO}}(\alpha)$, but the set

$$X = \{\beta : \beta \in V \text{ and } \mathfrak{V} \models [\varphi_{\mathrm{ORD}}(\beta) \wedge \beta < \alpha]\}$$

is not well-ordered. The set X is not well-ordered, but the model does not know about it. How is it possible? There is a $Y \subseteq X$ that has no least element, but the secret is, that no such Y can be an element of V. $\mathfrak{V}$ refuses to see that X is not-well-ordered.

8.3. Other Forms of Nonstandardness

The oddities of nonstandard models of arithmetic and set theory are related to the limitations of the expressive power of first-order logic. One cannot capture all properties of an infinite structure in a first-order way. The point of this chapter is to show that this is also a strength of first-order logic.

If $\mathfrak{M} \prec \mathfrak{N}$, then it would not always be proper to call $\mathfrak{N}$ a nonstandard extension of $\mathfrak{N}$. For example if $(D, <)$ is an elementary extension of $(\mathbb{Q}, <)$ and D is countable, then, by Cantor's theorem, $(D, <)$ is isomorphic to $(\mathbb{Q}, <)$. We assumed that elementary extensions are always proper, but the fact that new elements are added to the domain of the extension does not always make the extension a structure of a different kind.

For a given structure we are interested in elementary extensions with radically different properties. One way to achieve it is by iterations. For every structure we can form long chains of elementary extensions and in this way we can construct models of arbitrary high cardinalities. We will take advantage of it in Chapter 9, in the discussion of uncountable dense linear orderings. However, the cardinalirty of an extension alone does not make it automatically nonstandard. We are interested in extensions that alter the structure in a way that is useful for its model-theoretic analysis.

There are techniques that can only be applied to structures that are rich enough, in particular to models that admit nontrivial automorphisms. It turns out that in such cases, even if the structure we want to study is rigid, such as the standard model of arithmetic, or the field of real numbers, we can move to a (nonstandard) elementary extension that admits automorphisms, use the techniques there and then transfer the results back to learn something about the intended structure.

An active area of research where nonstandard extensions are routinely used is nonstandard analysis. Abraham Robinson, who was one of the two founding fathers of model theory (the other was Alfred Tarski), used model theory to formalize infinitesimal calculus. In [**Rob66**], he wrote

> ... Leibniz's ideas can be fully vindicated and ... they lead to a novel and fruitful approach to classical Analysis and to many other branches of mathematics. The key to our method is provided by the detailed analysis of the relation between mathematical languages and mathematical structures which lies at the bottom of contemporary model theory.

8.4. Nonstandard Versions of $(\mathbb{N}, <)$ and $(\mathbb{Z}, <)$

Recall that a structure is minimal, if all parametrically definable subsets of its domain are either finite or cofinite (Definition 3.9). In Chapter 3, we used automorphisms of $(\mathbb{Q}, <)$ to prove that this ordered set is minimal and the same arguments apply to $(\mathbb{R}, <)$. Here we will show that $(\mathbb{N}, <)$ is minimal, but the argument must be different, since $(\mathbb{N}, <)$ is rigid. For the proof we first need to take a closer look at elementary extensions of $(\mathbb{N}, <)$.

Let $(N, <)$ be a countable elementary extension of $(\mathbb{N}, <)$ and let c be a new "non-standard" element of N. How is c positioned with respect to the standard numbers in $\mathbb{N}$? Could c, for example, be less than 3? This cannot happen. That is because

$$(\mathbb{N}, <) \models \forall x[x < 3 \implies ((x = 0) \vee (x = 1) \vee (x = 2))]. \tag{$*$}$$

By elementarity, $(*)$ holds in $(N, <)$ and since c is a new element, it can't be less than 3. The same argument works for any other natural number. Being linearly ordered is a first-order property, so $(N, <)$ must be linearly ordered. Since we showed that for every natural number n, $(N, <) \models \neg(c < n)$, it follows that $(N, <) \models c > n$.

Now we will need the formula

$$\text{Succ}(x, y) = (x < y) \wedge \forall z[(x < z) \implies (y < z) \vee (y = z)]$$

that defines the successor relation in ordered sets and we will also use $\text{Pred}(x, y)$ defined as $\text{Succ}(y, x)$. We read $\text{Succ}(x, y)$ as "y is the successor of x" and $\text{Pred}(x, y)$ as "x is the predecessor of y." We have,

$$(\mathbb{N}, <) \models [\forall x \exists y \, \text{Succ}(x, y) \wedge (\forall y(y \neq 0 \implies \exists x \, \text{Pred}(x, y)))].$$

By elementarity, the same is true in $(N, <)$. This shows that together with a new element c, N must have infinitely many new elements that form a chain that is isomorphic to $(\mathbb{Z}, <)$. We will denote this chain by $\mathbb{Z}(c)$. There may be some nonstandard d in N that are not in $\mathbb{Z}(c)$, but one chain $\mathbb{Z}(c)$ is all we need to prove minimality of $(\mathbb{N}, <)$.

THEOREM 8.4. $(\mathbb{N}, <)$ *is minimal.*

PROOF. Since all natural numbers are definable in $(\mathbb{N}, <)$, for the proof it is enough to consider sets definable without parameters. We will do it only for notational convenience. The proof is the same for formulas with parameters.

The proof is by contradiction. Suppose that there is a formula $\varphi(x)$ such that the set $X = \{n : (\mathbb{N}, <) \models \varphi(n)\}$ is neither finite nor cofinite. Let $(N, <)$ be an elementary extension of $(\mathbb{N}, <)$. Then, we have

$$(\mathbb{N}, <) \models \forall x \exists y \exists z[x < y \wedge \text{Succ}(y, z) \wedge \varphi(y) \wedge \neg\varphi(z)].$$

By elementarity, the same is true in $(N, <)$; hence, there are nonstandard c and d such that

$$(N, <) \models [\text{Succ}(c, d) \wedge \varphi(c) \wedge \neg\varphi(d)]. \tag{$*$}$$

Let $f : N \longrightarrow N$ be

$$f(x) = \begin{cases} x + 1, & \text{if } x \in \mathbb{Z}(c), \\ x, & \text{otherwise,} \end{cases}$$

where $x + 1$ is the successor of x in the ordering of N.

Clearly, f is one-to-one, onto, and it preserves the ordering of N. Hence, f is an automorphism of $(N, <)$.

Since $f(c) = d$, we have

$$(N, <) \models [\varphi(c) \Longleftrightarrow \varphi(d)].$$

This contradicts $(*)$ and finishes the proof.   □

In the proof above, we have used a nonstandard element in an elementary extension of $(\mathbb{N}, <)$. In fact we could have used any model of $\text{Th}((\mathbb{N}, <))$ that is not isomorphic to $(\mathbb{N}, <)$. Let us see why.

DEFINITION 8.5. Let $(A, <, \dots)$ be an ordered structure. A subset I of A is an *initial segment* if for all a in I and b in A, if $b < a$, then b is in I. If $(I, <, \dots)$ is a substructure of $(A, <, \dots)$ and I is an initial segment of A, then we say that $(A, <, \dots)$ is an *end extension* of $(I, <, \dots)$.

PROPOSITION 8.6. Let $(A, <)$ be a model of $\text{Th}((\mathbb{N}, <))$. Then, there is an initial segment $I \subseteq A$ such that $(I, <)$ is isomorphic to $(\mathbb{N}, <)$. Moreover, $(I, <) \prec (A, <)$.

PROOF. Define $f : \mathbb{N} \longrightarrow A$ as follows. Let $f(0)$ be the least element of A, and for each $n > 0$, let $f(n)$ be the n-th successor of $f(0)$ in A. Since $(A, <)$ is elementary extension to $(\mathbb{N}, <)$, f is well defined for all n in $\mathbb{N}$. Clearly, $I = f(\mathbb{N})$ is an initial segment of A and $(I, <)$ is isomorphic to $(\mathbb{N}, <)$.

We will show that $(I, <) \prec (A, <)$. We will take advantage of the fact that every element n of $\mathbb{N}$ is definable, as the n-th successor of the least element. We will use the formula $\text{Succ}_n(x, y)$ expressing that y is the n-successor of x; see Definition 4.20. We have, $(\mathbb{N}, <) \models \varphi(n)$ iff $(\mathbb{N}, <) \models \forall x[\text{Succ}_n(0, x) \Longrightarrow \varphi(x)]$ iff $(A, <) \models \forall x[\text{Succ}_n(0, x) \Longrightarrow \varphi(x)]$ iff $(A, <) \models \varphi(f(n))$. The same argument works for sentences with parameters, proving that $(I, <) \prec (A, <)$.   □

By Proposition 8.6, we can say that every model of $\text{Th}((\mathbb{N}, <))$ is an elementary end extension of $(\mathbb{N}, <)$.

Before moving further, let us note two promised examples that we could not give earlier. The first is an example of a non-isolated type and the second is an example of two elementarily equivalent structures that are not isomorphic.

EXAMPLE 8.7. Let $(N, <)$ be an elementary extension of $(\mathbb{N}, <)$, and let c be a nonstandard element in N. We will show that $\text{tp}(c)$ is not isolated. Suppose that $(N, <) \models \varphi(c)$. Then, $(N, <) \models \exists x \varphi(x)$, by elementarity, there is a standard n such that $(\mathbb{N}, <) \models \varphi(n)$. Since n has its own unique type in $(\mathbb{N}, <)$ and therefore also in $(N, <)$, $\text{tp}(n) \neq \text{tp}(c)$. This shows that $\varphi(x)$ does not isolate $\text{tp}(c)$.

EXAMPLE 8.8. $(\mathbb{N}, <, \dots)$ is elementarily equivalent but not isomorphic to $(N, <, \dots)$. The dots indicate an expansion to a structure for any first-order language. The same holds for $(\mathbb{Z}, <, \dots)$ and any of its elementary extensions $(Z, <, \dots)$.

While being densely ordered and being discretely ordered are first-order properties, being well-ordered is not. There is no first-order sentence φ such that for any linearly ordered ordered set $(A, <)$, $(A, <)$ is well-ordered if and only if $(A, <) \models \varphi$. This follows from the results mentioned in our discussion of nonstandardness in set theory, but there is a particularly simple proof using $(\mathbb{N}, <)$ and its elementary extension $(N, <)$.

Suppose φ is a first-order sentence characterizing well-orderings. Then φ holds in $(\mathbb{N}, <)$ and, by elementarity, it must hold in $(N, <)$ as well, implying that $(N, <)$ is well-ordered, but that is a contradiction, because for any nonstandard c in N the chain $\mathbb{Z}(c)$ has no least element. Hence, there is no such φ.

DEFINITION 8.9. $\mathfrak{M}$ is *strongly minimal* if every structure that is elementarily equivalent is also minimal.

The field of complex numbers and most known minimal structures are strongly minimal, but $(\mathbb{N}, <)$ is not. Let $(N, <)$ be an elementary extension of $(\mathbb{N}, <)$, and let c be a nonstandard element in N. Then, the set defined in $(N, <)$ by $x < c$ is neither finite nor cofinite. $(\mathbb{N}, <)$ is the simplest example of a structure that is minimal but not strongly minimal. It is a difficult open problem whether there is an example that is a field.

$(\mathbb{Z}, <)$ is not minimal. For example, the set defined by $0 < x$ is neither finite not cofinitite. Still, the proof of minimality of $(\mathbb{N}, <)$ can be repeated almost verbatim to show that $(\mathbb{Z}, <)$ is o-minimal. To do this one needs to show that every parametrically definable subset of $\mathbb{Z}$ is a union of a finite number of intervals and points. To prove it by contradiction, assume that the set defined by a formula $\varphi(x, \bar{a})$, with $\bar{a} = a_1, \dots, a_n$ in $\mathbb{Z}$, is not of this form. Then either

$$(\mathbb{Z}, <) \models \forall x \exists y \exists z [x < y \land \mathrm{Succ}(y, z) \land \varphi(y, \bar{a}) \land \neg\varphi(z, \bar{a})],$$

or

$$(\mathbb{Z}, <) \models \forall x \exists y \exists z [y < x \land \mathrm{Succ}(y, z) \land \varphi(y, \bar{a}) \land \neg\varphi(z, \bar{a})],$$

and from this point on the proof is the same, except that the automorphism f has to be modified if the first disjunct does not hold.

THEOREM 8.10. $(\mathbb{Z}, <)$ *is o-minimal.*

One can classify completely elementary extensions of $(\mathbb{N}, <)$ and $(\mathbb{Z}, <)$. The reader is asked to do it in Exercise 13.12.

8.5. Models of Presburger Arithmetic

The domains of $(\mathbb{N}, <)$ and $(\mathbb{Z}, <)$ are made of numbers, but that was not essential. Any countable set would do. For that reason, it is not quite fair to call elements of $N \setminus \mathbb{N}$ or $Z \setminus \mathbb{Z}$ numbers. To talk about nonstandard numbers, we need to consider structures with arithmetic operations. In this section we will discuss addition.

Structures $(\mathbb{N}, +)$ and $(\mathbb{Z}, +)$ seem similar, but, perhaps surprisingly, their complete theories fall into two different model-theoretic categories. We will describe those categories in Chapter 15, for now let us see what makes the structures different. One obvious difference is that $(\mathbb{Z}, +)$ is a group and $(\mathbb{N}, +)$ is not, but what is more important from the model-theoretic perspective is that $(\mathbb{N}, +)$ is a richer structure, as we will now see.

$\mathrm{Th}((\mathbb{N}, +))$ is called Presburger arithmetic, to honor Mojżesz Presburger who gave a complete axiomatization of it in 1929. We will abbreviate $\mathrm{Th}((\mathbb{N}, +))$ by Pr.

For all natural numbers m and n, $m < n$ if and only if

$$(\mathbb{N}, +) \models \exists x[x \neq 0 \wedge m + x = n];$$

hence the ordering of $\mathbb{N}$ is definable in $(\mathbb{N}, +)$. It follows that the formula $\varphi(y, z) = \exists x[x \neq 0 \wedge y + x = z]$ defines a linear ordering of every model of Pr and this ordering has the property $\forall x, y, z[x < y \implies x + z < y + z]$.

In contrast, we saw in Section 4.4 that the ordering of $\mathbb{Z}$ is not definable in $(\mathbb{Z}, +)$. The proof was an easy application of the automorphism of $(\mathbb{Z}, +)$ defined by $f(x) = -x$. It can also be shown $<$ cannot be parametrically definable in $(\mathbb{Z}, +)$, but the proof is much harder. It is based on a characterization of subsets of $\mathbb{Z}^n$, $n > 0$, that are parametrically definable in $(\mathbb{Z}, +)$. See [**Rot00**, Section 15.2]. Notice that the relation defined by $\varphi(y, z)$ is $(\mathbb{Z}, +)$ is just $y \neq z$.

EXERCISE 8.11. Use the fact that $<$ is not parametrically definable in $(\mathbb{Z}, +)$ to show that $\mathbb{N}$ is not parametrically definable in $(\mathbb{Z}, +)$.

In Proposition 8.6, we showed that every model of $\mathrm{Th}((\mathbb{N}, <))$ is an elementary end extension of $(\mathbb{N}, <)$. The proof can be adapted to prove the same result for models of Pr.

PROPOSITION 8.12. Every model of Pr is an elementary end extension of $(\mathbb{N}, +)$.

EXERCISE 8.13. Prove Proposition 8.12.

8.5.1. Recursive Definitions. This subsection is a digression. The proof of Proposition 8.12 uses the facts that addition of natural numbers is determined by the successor relation $\mathrm{Succ}(x, y)$ and that $\mathrm{Succ}(x, y)$ is definable in $(\mathbb{N}, <)$. When we say that addition is determined, we mean that it is defined by recursion as follows.

The successor function $f(x) = x + 1$ is defined by $f(x) = y \iff \mathrm{Succ}(x, y)$. Then, for each natural number n, we define $n + 0 = n$ and for each k, $n + (k+1) =$

$(n + k) + 1$. Recursively, the successor function determines the value of $n + k$ for all n and k.

Assuming that addition has been defined already, multiplication can also be defined recursively by $n \cdot 0 = 0$ and $n \cdot (k + 1) = n \cdot k + n$.

Both $+$ and $\cdot$ are defined in $(\mathbb{N}, <)$ by recursion. However, as shown in examples 3.11 and 3.12, neither $(\mathbb{N}, +)$, nor $(\mathbb{N}, \cdot)$ are minimal. Since $(\mathbb{N}, <)$ is minimal, it follows that $+$ and $\cdot$ do not have first-order definitions in $(\mathbb{N}, <)$. In contrast, we will see in Chapter 12 how all such recursive definitions can be converted to first-order ones in the language of $(\mathbb{N}, +, \cdot)$.

EXERCISE 8.14. Let S be the successor relation on the natural numbers. Use an elementary extension of $(\mathbb{N}, S)$ to prove that the ordering of the natural numbers is not definable in $(\mathbb{N}, S)$.

Let us summarize. The main theme of this section is that while any set that is definable in $(\mathbb{N}, +, <)$ is already definable in $(\mathbb{N}, +)$, it is not so for the pair of structures $(\mathbb{Z}, +, <)$ and $(\mathbb{Z}, +)$. Expanding $(\mathbb{Z}, +)$ by inclusion of the ordering changes model-theoretic properties of the structure. The ordering plays a very special role here. Very recently, Gabriel Conant proved a result that sheds an interesting light on this situation.

THEOREM 8.15 ([**Con18**]). *Suppose that $A \subset \mathbb{Z}^n$ is definable in $(\mathbb{Z}, +, <)$. Then either A is already definable in $(\mathbb{Z}, +)$ or $<$ is definable in $(\mathbb{Z}, +, A)$.*

8.6. Z-groups

Models of $\text{Th}((\mathbb{Z}, +, <))$ are sometimes also called models of Presburger arithmetic, but not to confuse them with the models of $\text{Th}((\mathbb{N}, +))$, we will use the group-theoretic terminology and we will call them $\mathbb{Z}$-groups.

DEFINITION 8.16. A $\mathbb{Z}$-group is a structure for the language $\{+, <\}$ that is elementarily equivalent to $(\mathbb{Z}, +, <)$.

Let $\mathfrak{G} = (G, +, <)$ be a $\mathbb{Z}$-group, let $M = \{n : \mathfrak{G} \models 0 \leq n\}$, and let $\mathfrak{M} = (M, +, <)$. We will show that $\mathfrak{M}$ is a model of Pr. A small detail, $\leq$ is not in the language of $\mathfrak{G}$, but we use it as an abbreviation: $x \leq y$ means $(x < y) \vee (x = y)$.

For a first-order sentence φ of the language $\{+, <\}$, let φ^* be obtained from φ by replacing each occurrence of a subformula of φ of the form $\exists x \psi(x)$ by $\exists x[0 \leq x \wedge \psi(x)]$ and each occurrence a subformula of φ of the form $\forall x \psi(x)$ by $\forall x[0 \leq x \implies \psi(x)]$. In other words, φ is obtained by restricting the range of all quantifiers to the nonnegative numbers of $\mathfrak{G}$. It nan be easily checked that for each φ, $(\mathbb{N}, +, <) \models [\varphi \iff \varphi^*]$.

For each sentence φ, $(\mathbb{N}, +, <) \models \varphi$ iff $(\mathbb{Z}, +, <) \models \varphi^*$ iff $\mathfrak{G} \models \varphi^*$ iff $\mathfrak{M} \models \varphi$; hence $\mathfrak{M}$ is a model of Pr.

We showed that every $\mathbb{Z}$-group contains a model of Pr. Now we will show that a converse also holds. It is an exercise in first-order logic. We will define a $\mathbb{Z}$-group in every model of Pr. To do that, we will follow the construction of the integers from the natural numbers in a way that works not just for the standard model but for all nonstandard models of Pr as well.

Let $\mathfrak{M} = (M, +)$ be a model of Pr. Then

$$\mathbb{Z}(\mathfrak{M}) = \{(0, n) : n \in M\} \cup \{(1, n) : n \in M \setminus \{0\}\}$$

is a subset of M^2 that is definable in $\mathfrak{M}$. Think of $(0, n)$ as n and of $(1, n)$ as $-n$. In what follows, we will mimic the definitions of the ordering and addition of the integers (positive and negative signed numbers). Notice that in every model of Pr, if $m > n$ then $m - k$ is well-defined as the unique k such that $n + k = m$.

For (i, m), (j, n) in $\mathbb{Z}(\mathfrak{M})$, we define

$$(i, m) + (j, n) = \begin{cases} (0, m + n), & \text{if } i = j = 0, \\ (1, m + n), & \text{if } i = j = 1, \\ (0, m - n), & \text{if } i = 0, j = 1, \text{ and } m > n, \\ (1, n - m), & \text{if } i = 0, j = 1, \text{ and } n > m, \\ (1, m - n), & \text{if } i = 1, j = 0, \text{ and } m > n, \\ (1, n - m), & \text{if } i = 1, j = 0, \text{ and } n > m. \end{cases}$$

Also, we define $(i, m) < (j, n)$ to hold if and only if one of the following holds:

- $i = 1$ and $j = 0$;
- $i = j = 0$ and $m < n$;
- $i = j = 1$ and $n < m$.

The structure $(\mathbb{Z}(\mathfrak{M}), +, <)$, with $+$ and $<$ defined as above is a $\mathbb{Z}$-group. How do we know that? The point is that $(\mathbb{Z}(\mathfrak{M}), +, <)$ is defined in $\mathfrak{M}$ by the same first-order formulas that define an isomorphic copy of $(\mathbb{Z}, +, <)$ in $(\mathbb{N}, +)$. Because the domain and the relations of this copy of $(\mathbb{Z}, +, <)$ are definable in $(\mathbb{N}, +)$, each sentence in $\text{Th}((\mathbb{Z}, +, <))$ becomes a sentence of $\text{Th}((\mathbb{N}, +))$ after each occurrence of $<$ in it is replaced by a formula defining the ordering of pairs given above. Because $\mathfrak{M}$ is a model of $\text{Th}((\mathbb{N}, +))$, it follows that $(\mathbb{Z}(\mathfrak{M}), +, <)$ is a model of $\text{Th}((\mathbb{Z}, +, <))$.

A group $(G, +, <)$ is *ordered*, if for all x, y, and z in G, if $x < y$, then $x + z < y + z$. This is a first-order property and $(\mathbb{Z}, +, <)$ is an ordered group; hence each $\mathbb{Z}$-group is an ordered group.

What follows now is a series of exercises in elementary arithmetic; they are also exercises in elementary equivalence.

In $(\mathbb{Z}, +, <)$, every number is either even or odd. Formally,

$$(\mathbb{Z}, +, <) \models \forall x \exists y [x = y + y \lor x = y + y + 1].$$

Let $\mathfrak{G} = (G, +, <)$ be a nonstandard $\mathbb{Z}$-group. The remark above shows that every element of G is either even or odd, as defined by the first-order formulas above.

Every element of $\mathfrak{G}$ has an immediate successor and an immediate predecessor, hence $\mathfrak{G}$ contains a subgroup that is an isomorphic copy of $(\mathbb{Z}, +, <)$, we will identify this subgroup with $(\mathbb{Z}, +, <)$.

For $c \in G$, let $[c] = \{c + n : n \in \mathbb{Z}\}$.

For c and d in G, if $[c] \cap [d] \neq \varnothing$, then, for some integer m and n, $c + n = d + m$, then $d = c + (n - m)$, hence $[c] = [d]$. The argument is simple, but we need to check that its last step is valid. How do we know that $d = c + (m - d)$? One can argue as follows.

Subtraction is definable in $(\mathbb{Z}, +, <)$ i.e $n - m = k$ if an only if $n = k + m$, so we can refer to it in first-order statements. Moreover,

$$(\mathbb{Z}, +, <) \models \forall x, y, v, w [x + v = y + w \implies y = x + (v - w)] \qquad (*)$$

Since $\mathfrak{G}$ is elementarily equivalent to $(\mathbb{Z}, +, <)$, $(*)$ holds in $\mathfrak{G}$ and this makes the argument above legitimate.

Let $[G] = \{[c] : c \in G\}$. We define an ordering of $[G]$ by: $[c] < [d]$ if and only if $[c] \neq [d]$ and $c < d$.

PROPOSITION 8.17. $([G], <)$ is a densely ordered set without endpoints.

PROOF. Let c and d in G be positive, nonstandard, and even. Let $c = c' + c'$, $d = d' + d'$, and $e = c' + d'$. Notice that c' and d' are nonstandard. Moreover, $[c'] < [d']$, because otherwise, it is easy to show that $[c] = [d]$.

For any positive nonstandard a and all standard n, $a + n < a + a$. It follows that $[c'] < [c] < [c + c]$ and this shows that $([G], <)$ has no last element and that it has no least element above $[0]$.

Assume now that $[c] < [d]$. We will show that $[c] < [e] < [d]$. For each standard n, $c' + n < d'$, hence $c' + d' + n < d' + d' = d$ and this implies that $[e] < [d]$. Also, for each standard n, $c' + n < c' + c' < c' + d' = e$, proving that $[c] < [e]$.

Similar arguments applied to negative numbers show that $[G]$ has no least element and no largest element below $[0]$ and that the ordering is dense on the negative side of $[0]$ as well. □

8.7. Infinite Ramsey's Theorem

The argument showing that to confirm the twin primes conjecture it would be enough to prove that there is an elementary extension of the standard model of arithmetic with a pair of nonstandard twin primes has no real applications, but there are similar arguments that give simplified proofs of theorems in combinatorics. We will examine one special case.

For a set X, $[X]^n$ is the set of all n-element subsets of X. If X is linearly ordered, we identify $[X]^n$ with the set of all n-tuples $(x_1, \ldots, x_n)$ in X^n, such that $x_1 < \cdots < x_n$.

If X is infinite and K is finite, then for every function $f : X \longrightarrow K$, there is an infinite $H \subseteq X$ such that f restricted to H is constant; otherwise, X

would be a finite union of finite sets. This fundamental feature of infinity is known as the *infinite pigeonhole principle*. The next theorem is similar, but much less obvious.

THEOREM 8.18. *Let X be an infinite set. For every finite K, every $n > 0$ and every function $f : [X]^n \longrightarrow K$ there exists an infinite $H \subseteq X$ such that f restricted to $[H]^n$ is constant. H is called a* homogeneous set for f.

Instead of functions $f : [X]^n \longrightarrow K$, the theorem can be formulated in terms of partitions of $[X]^n$ into unions of finitely many disjoint subsets, or, in the case of $n = 2$, in terms of finite colorings of the edges of full infinite graphs.

Theorem 8.18 is due to Frank Ramsey (1903-1930). It has important applications in many areas of mathematics. We will need it in the next chapter. The proof below, for the case of $n = 2$, is essentially the usual proof, but with a model-theoretic twist that makes it a bit easier.

PROOF. We can assume that X is countable. If $X = \{x_0, x_1, x_2, \dots\}$ is an enumeration of X, then Ramsey's Theorem for X is a straightforward consequence of Ramsey's Theorem for the set of indices $\{0, 1, 2, \dots\}$, so we can also assume that $X = \mathbb{N}$ and that $K = \{0, \dots, k\}$ for some natural number k.

Let $f : [\mathbb{N}]^2 \longrightarrow \{0, \dots, k\}$ be given. Consider the structure $\mathfrak{M} = (\mathbb{N}, <, f)$ and its elementary extension $\mathfrak{N} = (N, <, g)$. Formally, f is a ternary relation, i.e., the graph of f, but we will refer to it using the function notation.

Let c be an element of $N \setminus \mathbb{N}$. As was shown in the proof of minimality of $(\mathbb{N}, <)$, for every natural number n, we must have $n < c$.

We have

$$\mathfrak{M} \models \forall x, y[x < y \implies \exists z[z \leq k \wedge f(x, y) = z],$$

and, because f is a function,

$$\mathfrak{M} \models \forall x, y, z_1, z_2[(f(x, y) = z_1 \wedge f(x, y) = z_2) \implies z_1 = z_2].$$

Since $\mathfrak{M} \prec \mathfrak{N}$, the same sentence holds in $\mathfrak{N}$ about g; hence, g is a function with domain $[N]^2$ and codomain K. Let $k_0 = g(0, c)$.

We will define an increasing sequence m_n of natural numbers by induction. Let $m_0 = 0$. Then, c witnesses that

$$\mathfrak{N} \models \exists x[m_0 < x \wedge g(m_0, x) = k_0]. \tag{0}$$

Hence, since m_0 and k_0 are in $\mathbb{N}$, we have

$$\mathfrak{M} \models \exists x[m_0 < x \wedge f(m_0, x) = k_0]. \tag{0'}$$

By (0'), there is m in $\mathbb{N}$ such that $f(m_0, m) = k_0$. Let m_1 be a least such m.

Now we repeat the argument as follows. Let $k_1 = g(m_1, c)$. Then, c witnesses that

$$\mathfrak{N} \models \exists x[m_1 < x \wedge g(m_0, x) = k_0 \wedge g(m_1, x) = k_1]. \tag{1}$$

Hence,
$$\mathfrak{M} \models \exists x[m_1 < x \wedge f(m_0, x) = k_0 \wedge f(m_1, x) = k_1]. \tag{1'}$$

By (1'), there is m in $\mathbb{N}$ such that $m_1 < m$, $f(m_0, m) = k_0$ and $f(m_1, m) = k_1$. Let m_2 be the least such m.

Inductively, assume that we have $m_0 < m_1 < \cdots < m_n$ in $\mathbb{N}$ such that for all $i < j \le n$, $f(m_i, m_j) = g(m_i, c) = k_i$. Let $k_n = g(m_n, c)$. Then, c witnesses that

$$\mathfrak{N} \models \exists x[m_n < x \bigwedge \{g(m_i, x) = k_i : i \le n\} \wedge g(m_n, x) = k_n]; \tag{n}$$

Hence

$$\mathfrak{M} \models \exists x[m_n < x \bigwedge \{f(m_i, x) = k_i : i \le n\} \wedge f(m_n, x) = k_n]; \tag{n'}$$

Let m_{n+1} be the least x in $\mathbb{N}$ for which (n') holds.

Let $B' = \{m_n : n \in \mathbb{N}\}$. By the construction, for all $i < j$,

$$f(m_i, m_j) = g(m_i, c).$$

Let $B_i = \{x \in B' : g(x, c) = i\}$, and let $B = \bigcup_{i < k} B_i$. For each i, f is constant on $[B_i]^2$ and at least one B_i is infinite. Let H be an infinite B_i. $\qquad\square$

In the proof above we reduced Ramsey's theorem for pairs to the pigeonhole principle. The proof can be modified to prove Ramsey's theorem for all exponents n by induction, reducing the problem of finding a homogeneous set for a partition of $[X]^{n+1}$ to the problem of finding one for a partition of $[B]^n$, for a suitably defined B.

The proof of Ramsey's theorem does not give us much information about the homogeneous set for a given function f. Everything depends on what values the function g assigns to the pairs (m, c), for standard m. There is no direct way to control those values. A variant of the proof using models of arithmetic yields more information. One can show that if the graph of the function $f : [\mathbb{N}]^n \longrightarrow K$, where K is finite, is definable in $(\mathbb{N}, +, \cdot)$, then there exists an infinite homogeneous set for f that is also definable in $(\mathbb{N}, +, \cdot)$. See Theorem 2.2.16 in [KS06].

CHAPTER 9

Categoricity

In the previous chapter, we have seen examples of countable structures $\mathfrak{M}$ and $\mathfrak{N}$ such that $\mathfrak{M} \prec \mathfrak{N}$ and $\mathfrak{M}$ is not isomorphic to $\mathfrak{N}$. This gives us examples of pairs of countable nonisomorphic models that are elementarily equivalent. Some theories have countable models $\mathfrak{M}$ such that any elementary extension of $\mathfrak{M}$ is not isomorphic to it. For many theories, each countable model has an elementary extension that is not isomorphic to it. While, by Theorem 6.4, any structure with a finite domain is uniquely determined by its first-order theory, in general, no set of first-order axioms can uniquely determine the isomorphism type of a structure with a countable domain. This feature of first-order logic is often seen as its weakness. However, in some rare cases, first-order properties can capture a countable structure completely. Such is the case of densely linearly ordered sets without end points. By Cantor's theorem (Theorem 5.11), there is just one such ordered set up to isomorphism. There are more examples and, as we will see in this chapter, they are all rather special.

9.1. Countable Categoricity

All theories in this chapter are assumed to be either in a finite or a countable language. When I say that a model is countable, I mean that its domain is countable, i.e., it is of cardinality $\aleph_0$.

DEFINITION 9.1. A theory is $\aleph_0$-*categorical*, if all of its countable models are isomorphic. A countable structure $\mathfrak{M}$ is $\aleph_0$-*categorical* if $\mathrm{Th}(\mathfrak{M})$ is $\aleph_0$-categorical.

So $(\mathbb{Q}, <)$ is $\aleph_0$-categorical and we only need a small finite part of $\mathrm{Th}(\mathbb{Q}, <)$ to arrive at this conclusion. Let DLO be the theory with the following axioms.

(1) Axioms for linear orderings. (Definition 1.5).
(2) $\forall x \forall y [x < y \implies \exists z (x < z \wedge z < y)]$ (density).
(3) $\forall x \exists y \exists z [y < x \wedge x < z]$ (no endpoints).

DLO is $\aleph_0$-categorical, but if we drop axiom (3), then the resulting theory DLO$^-$ is not.

EXERCISE 9.2. Prove that DLO$^-$ has exactly four nonisomorphic countable models.

DEFINITION 9.3. We say that a theory S *axiomatizes* a theory T, if for every φ in T, $S \models \varphi$. We say that T is *finitely axiomatizable*, if it is axiomatized by a finite theory.

PROPOSITION 9.4. DLO axiomatizes $\text{Th}((\mathbb{Q}, <))$.

PROOF. Suppose $\text{Th}((\mathbb{Q}, <)) \models \varphi$. We need to show that DLO $\models \varphi$. Let $(D, <)$ be a model of DLO, and suppose that $(D, <) \models \neg\varphi$. By Corollary 7.8 we can assume that D is countable. Then $(D, <)$ is isomorphic to $(\mathbb{Q}, <)$; hence $(D, <) \models \varphi$, contradiction.   □

For another example, the IEQ be the following theory in the language with one binary relation symbol E.

(1) Axioms stating that E is an equivalence relation:
    (a) $\forall x\, E(x, x)$ (reflexivity);
    (b) $\forall x, y[E(x, y) \Longrightarrow E(y, x)]$ (symmetry);
    (c) $\forall x, y, z[(E(x, y) \wedge E(y, z) \Longrightarrow E(x, z)]$, (transitivity).
(2) For all n, $\exists x_1, \ldots, x_n \bigwedge \{\neg E(x_i, x_j) : 0 < i < j \leq n\}$ (there are infinitely many equivalence classes);
(3) For all n, $\forall x \exists x_1, \ldots, x_n \bigwedge \{x_i \neq x_j : 0 < i < j \leq n\} \wedge \bigwedge \{E(x, x_i) : 0 < i \leq n\}$ (all equivalence classes are infinite).

If (M, E) is a model of IEQ, then E is an equivalence relation that has infinitely many equivalence classes and each of its equivalence classes is infinite. Here and below, we use E as a formal relation symbol and as an informal name of the equivalence relation $E \subseteq M^2$.

EXERCISE 9.5. Use the back-and-forth method to show that IEQ is $\aleph_0$-categorical.

Let $\mathfrak{M} = (M, E)$ be a model of IEQ. For each a in M, let $[a]$ be the equivalence class of a, i.e., the set $\{b : (a, b) \in E\}$. It is easy to check that if f is a permutation of M such that for each a in M, f restricted to $[a]$ is a permutation of $[a]$, then f an automorphism of $\mathfrak{M}$. The composition of such an automorphism with a permutation of the set of all equivalence classes is also an automorphism. In particular, for any a and b in M, there is a automorphism f such that $f(a) = b$; hence $\text{tp}^{\mathfrak{M}}(a) = \text{tp}^{\mathfrak{M}}(b)$. It follows that all elements of M have the same type. Moreover, the type that all elements share is isolated, for example by the formula $x = x$.

EXERCISE 9.6. Show that every automorphism $\mathfrak{M}$ must be of one of the forms described above.

EXERCISE 9.7. Show that $\text{Aut}(\mathfrak{M}) = 2^{\aleph_0}$.

EXERCISE 9.8. For each tuple $\bar{a} = a_1, \ldots, a_n$ in $M^{<\omega}$, Let

$$\varphi_{\bar{a}}(\bar{x}) = \bigwedge \{E(x_i, x_j) : (a_i, a_j) \in E\} \wedge \bigwedge \{\neg E(x_i, x_j) : (a_i, a_j) \notin E\}.$$

Show that $\varphi_{\bar{a}}(\bar{x})$ isolates $\text{tp}(\bar{a})$.

As we have seen, if $\mathfrak{M}$ is a countable model of either DLO or IEQ, then or each $n > 0$, $\mathfrak{M}$ realizes only finitely many complete n-types. Moreover, for each $\bar{a}$ in $M^{<\omega}$, $\text{tp}^{\mathfrak{M}}(\bar{a})$ is isolated. This generalizes to all $\aleph_0$-categorical structures.

THEOREM 9.9. *A structure $\mathfrak{M}$ is $\aleph_0$-categorical if and only if for each $n > 0$, the set of complete n-types realized in $\mathfrak{M}$ is finite. Moreover, if $\mathfrak{M}$ is $\aleph_0$-categorical, then for each $\bar{a}$ in $M^{<\omega}$, $\text{tp}(\bar{a})$ is isolated.*

Theorem 9.9 was proved in 1959 independently by Erwin Engeler, Czesław Ryll-Nardzewski, and Lars Svenonius. The proof is based on the omitting types theorem that is slightly more advanced than the material discussed in these lectures and will not be covered. For a proof, see Theorem 2.3.13 in [CK90].

Let $\mathfrak{M}$ be a countable model of either DLO or IEQ. We know that for all $\bar{a}$, $\bar{b}$ in $M^{<\omega}$, if $\text{tp}^{\mathfrak{M}}(\bar{a}) = \text{tp}^{\mathfrak{M}}(\bar{b})$, then there is an automorphism f such that $f(\bar{a}) = f(\bar{b})$.

EXERCISE 9.10. Let $\mathfrak{M}$ be a countable model either DLO or IEQ. Suppose that for $\bar{a}$, $\bar{b}$ in $M^{<\omega}$, $\text{tp}^{\mathfrak{M}}(\bar{a}) = \text{tp}^{\mathfrak{M}}(\bar{b})$. Show that for each a in M there is a b in M such that $\text{tp}^{\mathfrak{M}}(\bar{a}, a) = \text{tp}^{\mathfrak{M}}(\bar{b}, b)$.

In the next section, we will see that the result from the exercise above also generalizes to all $\aleph_0$-categorical structures.

9.2. Homogeneity

Theorem 9.9 implies that if $\mathfrak{M}$ is $\aleph_0$-categorical, then for each $n > 0$, there must be infinitely many n-tuples in $\mathfrak{M}^n$ that have the same type. Theorem 9.13 below will tell us how tuples of different length interact. First we need a definition and a lemma.

DEFINITION 9.11. A structure $\mathfrak{M}$ is *homogeneous* if for all $\bar{a}$, $\bar{b}$ in $M^{<\omega}$ if $\text{tp}(\bar{a}) = \text{tp}(\bar{b})$, then for each a in M there is b in M such that $\text{tp}(\bar{a}, a) = \text{tp}(\bar{b}, b)$. A structure $\mathfrak{M}$ is *strongly homogeneous* if for all $\bar{a}$, $\bar{b}$ in $M^{<\omega}$ if $\text{tp}(\bar{a}) = \text{tp}(\bar{b})$, then there is an automorphism f of $\mathfrak{M}$ such that $f(\bar{a}) = \bar{b}$.

LEMMA 9.12. *If for each $\bar{a}$ in $M^{<\omega}$, $\text{tp}^{\mathfrak{M}}(\bar{a})$ is isolated, then $\mathfrak{M}$ is homogeneous.*

PROOF. Suppose that for $\bar{a}$ and $\bar{b}$ in $M^{<\omega}$, $\text{tp}(\bar{a}) = \text{tp}(\bar{b})$, and let a in M be given. By the assumption, $\text{tp}(\bar{a}, a)$ is isolated. Let $\varphi(\bar{x}, y)$ be a formula that isolates it. Then

$$\mathfrak{M} \models \exists y \varphi(\bar{a}, y),$$

thus $\exists y \varphi(\bar{x}, y)$ is in $\text{tp}(\bar{a})$. Because $\bar{a}$ and $\bar{b}$ have the same type, $\exists y \varphi(\bar{x}, y)$ is in $\text{tp}(\bar{b})$. Hence,

$$\mathfrak{M} \models \exists y \varphi(\bar{b}, y).$$

Let b in M be such that $\mathfrak{M} \models \varphi(\bar{b}, b)$. Because $\varphi(\bar{x}, y)$ isolates $\text{tp}(\bar{a}, a)$, $\text{tp}(\bar{a}, a) = \text{tp}(\bar{b}, b)$, and this finishes the proof. $\square$

In Lemma 9.12 we have not assumed that M is countable, if it is, we can prove more.

THEOREM 9.13. *Countable homogeneous structures are strongly homogeneous.*

PROOF. The proof is modeled on the proof of Theorem 5.11. Let $\mathfrak{M}$ be countable, homogeneous, and assume that $\bar{a}$ and $\bar{b}$ in $M^{<\omega}$ have the same type. Let $\{c_n : n \in \mathbb{N}\}$ be an enumeration of M. Let $d_0 = c_0$. By Lemma 9.12, there is a e_0 such that $\text{tp}(\bar{a}, d_0) = \text{tp}(\bar{b}, e_0)$. This completes the first "forth" step in the construction. Now let e_1 be the first element among $\{c_n : n \in \mathbb{N}\}$ that has not been used in the construction yet. Again, by homogeneity, there is an d_1 such that $\text{tp}(\bar{a}, d_0, d_1) = \text{tp}(\bar{b}, e_0, e_1)$. This completes the first "back" step.

Now we proceed this way to inductively define two sequences $\{d_n : n \in \mathbb{N}\}$ and $\{e_n : n \in \mathbb{N}\}$ such that for each n, $\text{tp}(\bar{a}, d_0, \ldots, d_n) = \text{tp}(\bar{b}, e_0, \ldots, e_n)$ and each element of M appears on both lists. It is straightforward to check that the function defined by $\bar{a} \mapsto \bar{b}$ and $d_n \mapsto e_n$, for all n, is an automorphism of $\mathfrak{M}$. $\qquad\square$

Because all countable homogeneous structures are strongly homogeneous, for examples separating these two notions, we have to turn to uncountable structures. One example is given in the next exercise.

EXERCISE 9.14. Let $\mathbb{Q}^-$ be the set of negative rational numbers, and let $\mathbb{R}^+$ be the set of positive real numbers. Show that $(\mathbb{Q}^- \cup \mathbb{R}^+, <)$ is homogeneous, but not strongly homogeneous.

The following corollary of Theorem 9.13 follows directly from Theorem 9.9.

COROLLARY 9.15. $\aleph_0$-categorical structures are strongly homogeneous.

Not all countable strongly homogeneous structures are $\aleph_0$-categorical. We will see interesting examples in Chapter 13. For now we have an uninteresting example in the following exercise.

EXERCISE 9.16. Show that $(\mathbb{N}, <)$ is strongly homogeneous. HINT: It follows immediately from the definition of strong homogeneity.

As an application of homogeneity, we will prove the following useful corollary. The reader is encouraged to supply a proof before reading the proof below. Recall that for $\bar{a}$ in $M^{<\omega}$, the language of the expanded structure $(\mathfrak{M}, \bar{a})$ includes new constant symbols for the elements in the tuple $\bar{a}$. Then, we can say that for some $\bar{b}$ in $N^{<\omega}$, $(\mathfrak{N}, \bar{b})$ is a model of $\text{Th}((\mathfrak{M}, \bar{b}))$, if the corresponding elements in $\bar{b}$ are interpretations of those constant symbols in $\mathfrak{N}$.

COROLLARY 9.17. If $\mathfrak{M}$ is $\aleph_0$-categorical, then so is $(\mathfrak{M}, \bar{a})$, for all $\bar{a}$ in $M^{<\omega}$.

PROOF. Let $(\mathfrak{N}, \bar{b})$ be a countable model of $\text{Th}(\mathfrak{M}, \bar{a})$. Since $\mathfrak{N} \models \text{Th}(\mathfrak{M})$, $\mathfrak{N}$ is isomorphic to $\mathfrak{M}$. In particular, $\mathfrak{N}$ is strongly homogeneous. Let $f :$

$M \longrightarrow N$ be an isomorphism, and let $\bar{c} = f(\bar{a})$. Then $\mathrm{tp}(\bar{a}) = \mathrm{tp}(\bar{b}) = \mathrm{tp}(\bar{c})$. By Theorem 9.13, there is an g in $\mathrm{Aut}(\mathfrak{N})$ such that $g(\bar{c}) = \bar{b}$. Then $g \circ f(\bar{a}) = \bar{b}$, showing that $(\mathfrak{M}, \bar{a})$ and $(\mathfrak{N}, \bar{b})$ are isomorphic. $\qquad\square$

In contrast to the above corollary, for each countable $\mathfrak{M}$, $\mathrm{Th}(\mathfrak{M}, a)_{a \in M}$ is not $\aleph_0$-categorical. Indeed, if $(\mathfrak{N}, a)_{a \in M}$ is a proper elementary extension of $(\mathfrak{M}, a)_{a \in M}$, then no b in $N \setminus M$ is interpreted as a constant from M; hence $(\mathfrak{M}, a)_{a \in M}$ and $(\mathfrak{N}, a)_{a \in M}$ are not isomorphic.

9.3. The Random Graph

The countable random graph is a rather curious object. Instead of defining it directly, we will first define its theory.

DEFINITION 9.18. Let RAND be the following set of axioms in the language of graphs.

$$\forall x, y[\mathcal{E}(x, y) \implies \mathcal{E}(y, x)],$$

$$\forall x \neg \mathcal{E}(x, x),$$

and for all n,

$$\forall x_1, \dots, \forall x_n, \forall y_1, \dots, y_n[\bigwedge \{x_i \neq y_j : 1 \leq i, j \leq n\} \implies$$

$$\exists z \bigwedge \{\mathcal{E}(x_i, z) \wedge \neg \mathcal{E}(y_j, z) : 1 \leq i \leq n\}\}].$$

A graph with one vertex and no edges trivially is a model of RAND, but it is not clear at all if RAND has other models. If you are familiar with axiomatic set theory, then you may try the following exercise.

EXERCISE 9.19. Let $\mathfrak{V} = (V, \in)$ be a countable model of ZFC, and let

$$E = \{(v, w) \in V^2 : (v \in w) \vee (w \in v)\}.$$

Show that $(V, E) \models$ RAND. HINT: For disjoint sets $\{v_1, \dots, v_m\}$ and $\{w_1, \dots, w_n\}$ in V, consider $u = \{w_1, \dots, w_n\}$ and $z = \{v_1, \dots, v_n, u\}$.

For a more direct construction of a countable model of RAND see [**Mar02**, Theorem 2.4.2].

EXERCISE 9.20. Use a back-and-forth argument to prove that RAND is $\aleph_0$-categorical.

Because RAND is $\aleph_0$-categorical, it justifies calling any of its models the *random graph*. The countable random graph is also called the *Rado graph*.

9.4. Completeness of Theories

Up to now we have considered a theory T to be complete if for every sentence of its language either φ or $\neg\varphi$ is in T. Now we will switch to a more general definition.

DEFINITION 9.21. A theory T is complete if for every sentence φ in $\mathcal{L}_T$ either $T \models \varphi$ or $T \models \neg\varphi$.

PROPOSITION 9.22. $\aleph_0$-categorical theories are complete.

PROOF. If T is inconsistent, then for all φ, $T \models \varphi$, so it is complete.

Let T be a consistent $\aleph_0$-categorical theory. By Corollary 7.8, T has a countable model $\mathfrak{M}$. We will show that for all φ, $T \models \varphi$ if and only if $\mathfrak{M} \models \varphi$. Because $\mathrm{Th}(\mathfrak{M})$ is complete, this will finish the proof.

Suppose that $T \models \varphi$. This means that φ is true in all models of T, hence $\mathfrak{M} \models \varphi$.

For the other direction, assume that $\mathfrak{M} \models \varphi$ and, to get a contradiction, suppose that φ is not a consequence of T. It means that there is a model $\mathfrak{N}$ of T such that $\mathfrak{N} \models \neg\varphi$. By Corollary 7.8, we can assume that N is countable, so $\mathfrak{N}$ is isomorphic to $\mathfrak{M}$; hence $\mathfrak{M} \models \neg\varphi$. Contradiction. □

COROLLARY 9.23. DLO, IEQ and RAND are complete.

In Definition 9.1, we might have assumed that the theory is complete. If T is incomplete, then there is a sentence φ and two models $\mathfrak{M}$ and $\mathfrak{N}$ of T such that $\mathfrak{M} \models \varphi$ and $\mathfrak{N} \models \neg\varphi$, so $\mathfrak{M}$ and $\mathfrak{N}$ are not isomorphic. By Corollary 7.8, we can assume that M and N are countable; hence T is not $\aleph_0$-categorical.

Investigations of completeness of theories is an important task of model theory. Some theories are incomplete for the simple reason that they are not specific enough. For example, the axioms of dense linear orderings are not a complete theory because they do not determine whether there are least or last elements in the ordering, and those properties are first-order expressible in the language of orderings. Because DLO is complete, these are the only properties that those axioms do not decide.

The group axioms are not complete on purpose, it is a minimal set of axioms that captures the largest possible class of groups, so, for example, the axioms do not determine if the group operation is commutative or not.

There are also theories intended to capture properties of a single structure, such as the standard model of arithmetic. In such cases we want them to be as complete as possible. It turns out that most such theories are incomplete due to Gödel's incompleteness theorem that says that for each effectively presented (computable) consistent set axioms that are rich enough, there are sentences that are undecided by the axioms. See [**Kay91**, Chapter 3] for a precise statement and a proof of Gödel's theorem.

9.5. Uncountable Categoricity

In these lectures, the use of set theory is limited to a necessary minimum. This allows us to discuss fully some restricted, but still interesting and applicable results that in their general formulations use more advanced set theory. A disadvantage is that some very attractive topics cannot be fully covered. Such is the general case of categoricity that will be outlined briefly in this section.

If a theory has a model with an infinite domain, then, by Theorem 7.10, it has models with domains of arbitrarily large cardinalities. In this sense, no such theory can be fully categorical, but there is a very meaningful localized notion of categoricity that applies to all theories.

For a cardinal number κ, a theory T is κ-categorical if any two of its models with domains of cardinality κ are isomorphic. In 1965, Michael Morley confirmed the Łoś conjecture: if a theory in a countable language is κ-categorical for some uncountable κ, then it is κ-categorical for all uncountable κ.

Every complete theory T in a countable language that has models with infinite domains falls into one of the four categories below.

- T is κ-categorical for all infinite cardinals κ. Such theories are called totally categorical. There are very few natural examples of totally categorical theories. One is the theory of infinite vector spaces over a finite field. Every such space of cardinality κ has a basis of cardinality κ; hence any two such spaces are isomorphic.

- T is κ-categorical only for $\kappa = \aleph_0$. It is easy to show that DLO, IEQ and RAND are not κ-categorical for any uncountable κ and that is because uncountable models of these theories can have parts of different cardinalities. For example, $(\mathbb{R}, <)$ and $(\mathbb{R} \oplus \mathbb{Q}, <)$, where $\mathbb{R} \oplus \mathbb{Q}$ is $\mathbb{R}$ with a copy of $\mathbb{Q}$ appended at the top, are nonisomorphic models of DLO of cardinality $2^{\aleph_0}$. A more elaborate set-theoretic construction below shows that for every regular cardinal κ, there are 2^κ pairwise nonisomorphic models of DLO of cardinality κ.

- T is κ-categorical if and only if $\kappa > \aleph_0$. An important example of a theory that is κ-categorical for all uncountable κ, but not $\aleph_0$-categorical, is the theory of the complex field $\text{Th}((\mathbb{C}, +.\cdot))$. See [**Mar02**, Corollary 3.2.9].

- T is κ-categorical for no infinite cardinal κ. Examples of such theories are $\text{Th}((Z, +))$ and $\text{Th}((\mathbb{N}, +, \cdot))$. We will see why in Chapter 12.

EXERCISE 9.24. Show that the theory of vector spaces over the field of rational numbers is not ω-categorical, but is κ-categorical for each uncountable cardinal number κ.

9.5.1. Long Elementary Chains. In this section, we will see how an $\aleph_0$-categorical theory can fail to be $\aleph_1$-categorical. This material is more advanced. The aim is to show how set theory brings complexity into model theory.

Let $\mathfrak{M}$ be $\aleph_0$-categorical. By Corollary 7.10, $\mathfrak{M}$ has a countable elementary extension $\mathfrak{M}_1$ and because $\mathfrak{M}$ is $\aleph_0$-categorical, $\mathfrak{M}_1$ is isomorphic to $\mathfrak{M}$. Now

$\mathfrak{M}_1$ also has a countable elementary extension $\mathfrak{M}_2$ that is isomorphic to $\mathfrak{M}$. This can continued: for every countable ordinal α, we let $\mathfrak{M}_{\alpha+1}$ be an elementary extension of $\mathfrak{M}_\alpha$, and for countable limit ordinal λ, we let $\mathfrak{M}_\lambda$ be the union of the elementary chain $\{M_\alpha\}_{\alpha<\lambda}$. By the elementary chain lemma, $\mathfrak{M}_\lambda$ is an elementary extension of each $\mathfrak{M}_\alpha$ in the chain and, because it is countable, $\mathfrak{M}_\lambda$ is isomorphic to $\mathfrak{M}$.

Suppose that $\mathfrak{M}$ has two countable elementary extensions $\mathfrak{M}_1$ and $\mathfrak{M}'_1$ such that the structures $(\mathfrak{M}_1, M)$ and $(\mathfrak{M}'_1, M)$, for the language of $\mathfrak{M}$ extended with a new unary relation symbol interpreted as M, are not isomorphic. In other words, while $\mathfrak{M}_1$ and $\mathfrak{M}_2$ are isomorphic, M is positioned inside them in two different ways, making the expanded structures nonisomorphic. Let us call $\mathfrak{M}_1$ extension of type A and $\mathfrak{M}'_1$ extension of type B. Moreover, let us assume that the extensions of the given type can be preserved by further extensions. This means that we can build the chains $\{\mathfrak{M}_\alpha\}_{\alpha<\aleph_1}$ so that not only $(\mathfrak{M}_{\alpha+1}, M_\alpha)$ is of a given type, but also $(\mathfrak{M}_\beta, M_\alpha)$, for all $\beta > \alpha$, are of that type. We will see examples of such extensions shortly.

Let $\mathfrak{N}_A$ and $\mathfrak{N}_B$ be the unions of elementary chain extensions as above, such that the extensions are always of type A in the chain of $\mathfrak{N}_A$ and of type B in the chain of $\mathfrak{N}_B$. We will show that then $\mathfrak{N}_A$ and $\mathfrak{N}_B$ are not isomorphic. This will prove that $\mathrm{Th}(\mathfrak{M})$ is not $\aleph_1$-categorical.

The last claim is a consequence of the following general result. For an ordinal γ, we say that $\{\mathfrak{M}_\alpha\}_{\alpha<\gamma}$ is a *continuous elementary chain* if for all $\alpha < \gamma$, $\mathfrak{M}_\alpha \prec M_{\alpha+1}$, and for limit λ, $\mathfrak{M}_\lambda$ is the union of the chain $\{\mathfrak{M}_\alpha\}_{\alpha<\lambda}$.

Each cardinal number κ is the set of all ordinal numbers α such that $\alpha < \kappa$. A set $X \subseteq \kappa$ is *unbounded* if for every $\alpha < \kappa$, there is a $\beta \in X$ such that $\alpha < \beta$; X is *closed* if for each each for each bounded sequence $\{\alpha_\beta\}_{\beta<\lambda}$ of elements of X, where λ is a limit ordinal, $\sup\{\alpha_\beta\}_{\beta<\lambda}$ is in X.

PROPOSITION 9.25. Let κ be a regular uncountable cardinal, and let $\mathfrak{M}$ and $\mathfrak{N}$ be the unions of continuous elementary chains $\{\mathfrak{M}_\alpha\}_{\alpha<\kappa}$ and $\{\mathfrak{N}_\alpha\}_{\alpha<\kappa}$ such that for all $\alpha < \kappa$, $|M_\alpha| < \kappa$ and $|N_\alpha| < \kappa$. If $f : M \longrightarrow N$ is an isomorphism, then $\{\alpha < \kappa : f(M_\alpha) = N_\alpha\}$ is closed and unbounded in κ.

Proposition 9.25 can be proved by a standard Löwenheim-Skolem argument. With a bit more set theory, we can prove more.

A subset X of κ is *stationary* if it has a nonempty intersection with each closed and unbounded subset of κ. It can be shown that every uncountable regular cardinal number κ can be partitioned into κ pairwise disjoint stationary subsets. For each set X of those subsets, we can define $\mathfrak{M}_X$ as the union of a chain $\{\mathfrak{M}_\alpha\}_{\alpha<\kappa}$ of extensions of type A for $\alpha \in X$ and type B for $\alpha \notin X$. Then, it follows from Proposition 9.25 that if $X \neq Y$, then $\mathfrak{M}_X$ and $\mathfrak{M}_Y$ are not isomorphic. Hence, the theory of those models is as not κ-categorical as it can be. It has 2^κ pairwise nonisomorphic models of cardinality κ.

We will apply this construction to get many nonisomorphic dense linear orderings.

9.5.2. Noncategoricity of DLO. For ordered sets $(A, <)$, $(B, <)$, $(A \oplus B, <)$ is the ordered set whose domain is the disjoint union of A and B, the ordering on copies of A and B in that union is inherited from A and B, respectively, and $a < b$ for all a in A and b in B.

Let $(Q_0, <)$ and $(P_0, <)$ both be $(\mathbb{Q}, <)$, and let $\mathbb{P}$ be the set of nonnegative rationals. It is important that $\mathbb{P}$ has a least element. Now we proceed by transfinite recursion.

For each ordinal number α, we let $(Q_{\alpha+1}, <)$ be $(Q_\alpha \oplus \mathbb{Q}, <)$ and $(P_{\alpha+1}, <)$ be $(P_\alpha \oplus \mathbb{P}, <)$. For limit λ, we take the unions, $(Q_\lambda, <) = (\bigcup_{\alpha < \lambda} Q_\alpha, <)$ and $P_\lambda = (\bigcup_{\alpha < \lambda} P_\alpha, <)$.

For all $\alpha < \beta$, $(Q_\beta, Q_\alpha, <)$ and $(P_\beta, P_\alpha, <)$ are not isomorphic. The set $P_\beta \setminus P_\alpha$ has a least element, but $Q_\beta \setminus Q_\alpha$ does not. So we have an example of elementary extensions of two types A and B, as discussed in the previous section.

Let κ be an uncountable regular cardinal. Then $|Q_\kappa| = |P_\kappa| = \kappa$ and, by Proposition 9.25, $(Q_\kappa, <)$ is not isomorphic to $(P_\kappa, <)$. Using the construction from the previous section, we get 2^κ pairwise nonisomorphic dense linear orderings of cardinality κ.

9.6. Truth in Algebraically Closed Fields

Unlike most of the material in this book, this section is not self-contained. Nevertheless, it is included as the results are a prime example of early important applications of model theory in algebra. For a fuller account, the reader is referred to [**Mar02**, Section 3.1]. We will follow Marker's exposition.

The first-order theory of algebraically closed fields, ACF, consists of the field axioms and the axiom schema:

$$\forall y_0, \ldots, y_{n-1} \exists x \; x^n + \sum_{i=0}^{n-1} y_i x^i = 0,$$

for all $n > 0$. ACF is not a complete theory. It does not decide the characteristic of the field. For a prime number p, let χ_p be the following sentence expressing that the charecteristic of a field is p:

$$\forall x \underbrace{x + \cdots + x}_{p-\text{times}} = 0.$$

The field of complex numbers $\mathfrak{C}$ is a model of ACF_0.

ACF_p is $\mathsf{ACF} + \chi_p$ and ACF_0 is $\mathsf{ACF} + \{\neg\chi_p : p \text{ prime }\}$. For $p = 0$ and for all prime p, there is a unique model of ACF_p of cardinality $2^{\aleph_0}$; hence each ACF_p is complete. From this we get the following corollary.

COROLLARY 9.26. For each sentence φ of the language $\{+, \cdot\}$, the following are equivalent:

(1) $\mathfrak{C} \models \varphi$.
(2) $\mathfrak{M} \models \varphi$, for every $\mathfrak{M} \models \mathsf{ACF}_0$.

(3) $\mathfrak{M} \models \varphi$, for some $\mathfrak{M} \models \mathsf{ACF}_0$.

(4) There are arbitrarily large p such that $\mathfrak{M} \models \varphi$ for some $\mathfrak{M} \models \mathsf{ACF}_p$.

(5) There is m such that for all prime $p > m$, $\mathfrak{M} \models \varphi$ for all $\mathfrak{M} \models \mathsf{ACF}_p$.

PROOF. (1), (2), (3) follow from the completeness of ACF_0 and clearly (5) implies (4). We will show that (2) implies (5). Assume (2), which means that $\mathsf{ACF}_0 \models \varphi$. Then, by the Finiteness Theorem (Theorem 7.11), there is a finite $S \subseteq \mathsf{ACF}_0$, such that $S \models \varphi$. Let m be the largest q such that χ_q is in S. Because for each prime $p > m$, $\mathsf{ACF}_p \models \chi_q$, for all $q \leq m$, and this gives us (5). □

We will use Corollary 9.26 to prove the following purely algebraic result. The model-theoretic proof below, due to James Ax, was the original proof of the result. The result was later given algebraic proofs.

THEOREM 9.27. *Every one-to-one polynomial map from $\mathbb{C}^n$ from $\mathbb{C}^n$ is onto.*

PROOF. We will only give a brief sketch of the proof. For more details see [**Mar02**, Theorem 2.2.11]. For a field F, let $\bar{F}$ be the algebraic closure of F, and for each prime p, let $\mathbb{F}_p$ be the unique p-element field.

For each n and d the statement of the theorem for all polynomial maps of degree d can be formalized as a single first-order sentence φ of the language $\{+, \cdot\}$. By Corollary 9.26, to prove that $\mathfrak{C} \models \varphi$ it suffices to prove that it holds in all fields $\overline{\mathbb{F}_p}$. So let K be $\overline{\mathbb{F}_p}$, and let $f : K^n \longrightarrow K^n$ be a polynomial map of degree d. Let $\bar{b}$ in K^n be given, and let L be the smallest subfield of K that contains all elements in the n-tuple $\bar{b}$ and all coefficients of f. Now we use the fact that each finite subset of K is contained in a finite subfield of K. Hence L is finite and f restricted to L^n is a map from L^n to L^n. Because f is one-to-one, the image of L^n under f is L^n, implying that $\bar{b}$ is in the range of f and finishing the proof. □

CHAPTER 10

Indiscernibility

In this section we enter more advanced topics in model theory.

Indiscernibility is a powerful model-theoretic concept, but it is not easy to motivate at this point because its power only reveals itself in applications to problems that we have not had a chance to discuss yet. For those applications, we will be interested in structures that are generated by large ordered sets in which every increasing tuple of elements looks like any other increasing tuple of the same length. The idea of a set whose all finite sequences look alike is captured in the following definition. What it means for a structure to be generated by a set will be explained in the last section of this chapter.

DEFINITION 10.1. Let $(I, <)$ be a linearly ordered set. We say that a sequence $\{a_i : i \in I\}$ of elements of the domain of a structure $\mathfrak{M}$ is *indiscernible* if for all finite sequences $i_1 < \cdots < i_n$ and $j_1 < \cdots < j_n$ in I, $\mathrm{tp}^{\mathfrak{M}}(a_{i_1}, \ldots, a_{i_n}) = \mathrm{tp}^{\mathfrak{M}}(a_{j_1}, \ldots, a_{j_n})$.

In Definition 10.1, the ordering does not have to be a part of the structure $\mathfrak{M}$, although in many interesting cases it is. If $\mathfrak{M}$ is an ordered structure then we will tacitly assume that the ordering of $\mathfrak{M}$ serves as the ordering in Definition 10.1.

In any structure all one element subsets of the domain are indiscernible for trivial reasons and sometimes this is as much as we can get. For example, in $(\mathbb{N}, <)$ every element has its own unique type, hence no two element set can form an indiscernible sequence.

All integers have the same type in $(\mathbb{Z}, <)$; hence every two element sequence is indiscernible, but if a, b, and c are distinct integers, then $\mathrm{tp}(a, b) \neq \mathrm{tp}(a, c)$, as the distances between a and b, and a and c, are different, and since that can be expressed in a first-order way, in $(\mathbb{Z}, <)$ there are no indiscernible sequences of length 3.

We have seen in Section 5.3.2 that if $\bar{p} = p_1, \ldots, p_n$ and $\bar{q} = q_1, \ldots, q_n$ are increasing sequences of rational numbers, then the types of $\bar{p}$ and $\bar{q}$ in $(\mathbb{Q}, <)$ are the same. The types of finite sequences of rationals are determined by the ordering of the numbers in the sequences. This means that every finite increasing sequence of rational numbers is indiscernible. It follows that if $(I, <)$ is a countable ordered set and $A = \{p_i : i \in I\}$ is a sequence or rational numbers such that for all i and j, $p_i < p_j$ if and only if $i < j$, then A is an indiscernible sequence in $(\mathbb{Q}, <)$.

The remark above and Exercise 8.17 can be used to do the following exercise.

EXERCISE 10.2. Let $(\mathbb{Z}^*, +, <)$ be a $\mathbb{Z}$-group that is not isomorphic to $(\mathbb{Z}, +, <)$. Show that for every countable ordered set $(I, <)$, $(\mathbb{Z}^*, <)$ has an infinite indiscernible sequence $\{a_i : i \in I\}$.

10.1. Models with Indiscernibles

In Exercise 10.2, we get sequences that are indiscernible in the ordered structure $(\mathbb{Z}^*, <)$. There are also such sequences that are indiscernible in $(\mathbb{Z}^*, +, <)$ for some $\mathbb{Z}$-group $(\mathbb{Z}^*, +, <)$. Instead of proving it directly, we will prove a general theorem.

THEOREM 10.3. Let T be a theory, let $\mathfrak{M}$ be a model of T with an infinite domain. Then, for every ordered set $(I, <)$, $\mathfrak{M}$ has an elementary extension with an indiscernible sequence $\{a_i : i \in I\}$.

PROOF. Let T be a theory, and let $\mathfrak{M}$ be a model of T with an infinite domain. For a given ordered set $(I, <)$, we expand the language of T by adding constant symbols c_i for each $i \in I$. If $\mathfrak{M}$ is not linearly ordered by one of its relations, we also add the binary relation symbol $<$ to the language.

Consider the theory T_I consisting of the following axioms:

(1) $\text{Th}(M, a)_{a \in M}$;
(2) $\varphi(c_{i_1}, \ldots, c_{i_n}) \iff \varphi(c_{j_1}, \ldots, c_{j_n})$, for all finite sequences $i_1 < \cdots < i_n$ and $j_1 < \cdots < j_n$ in I, and all formulas $\varphi(x_1, \ldots, x_n)$ of the language of T;
(3) $c_i < c_j \iff i < j$, for all i and j in I.

We claim that T_I is consistent. Let T' be a finite subset of T_I. Let $\varphi_1, \ldots, \varphi_k$ be a list of the formulas occurring in the instances of (2) in T'. Let us consider $\varphi_1(x_1, \ldots, x_n)$ first. Let $f_1 : [M]^n \longrightarrow \{0, 1\}$ be defined by

$$f_1(x_1, \ldots, x_n) = \begin{cases} 1, & \text{if } \mathfrak{M} \models \varphi_1(x_1, \ldots, x_n), \\ 0, & \text{otherwise.} \end{cases}$$

By Ramsey's theorem (Theorem 8.18), there is an infinite $H_1 \subseteq M$ that is homogeneous for f_1. We can now interpret all constants c_i occurring in T' as terms of a finite increasing sequence of elements of H_1, so that all instances of (2) and (3) in T' will hold.

We repeat the argument for $\varphi_2, \ldots, \varphi_k$, to obtain a nested sequence of infinite sets $H_1 \supseteq H_2 \supseteq \cdots \supseteq H_k$, such that for all i, H_i is homogeneous for φ_i. Interpreting the constants in T' by terms of an increasing finite sequence of elements of H_k we get a model of T'. By the compactness theorem, this finishes the proof. $\square$

10.2. The Ehrenfeucht Mostowski Theorem

Theorem 10.3 was proved by Andrzej Ehrenfeucht and Andrzej Mostowski in [**EM56**]. One of the questions that was asked in the early days of model theory was whether there could be first-order theories with infinite models all of whose models are rigid. Ehrenfeucht and Mostowski used their theorem to show that there are no such theories. In this section we will examine their proof in a special case.

For the rest of this section, let T be a theory in a language $\mathcal{L}_T$ that includes a binary relation symbol $<$. For any formula $\varphi(x, \bar{y})$ of $\mathcal{L}_T$, let $\mu_\varphi(x, \bar{y})$ be

$$\varphi(x, \bar{y}) \wedge \forall z[z < x \implies \neg\varphi(z, \bar{y})].$$

We assume that T proves that $<$ is a linear ordering and that it proves all sentences of the form:

$$\forall \bar{y}[\exists x \varphi(x, \bar{y}) \implies \exists x \mu_\varphi(x, \bar{y})]. \tag{$*$}$$

Examples of such T are Peano Arithmetic and $\mathrm{Th}(\mathbb{N}, <, \dots)$, where $(\mathbb{N}, <, \dots)$ is any expansion of $(\mathbb{N}, <)$.

DEFINITION 10.4. Let $\mathfrak{M}$ be a model of T. For each $A \subseteq M$, the *Skolem closure* of A in $\mathfrak{M}$, denoted $\mathrm{Scl}^{\mathfrak{M}}(A)$, is the set of all b such that $\mathfrak{M} \models \mu_\varphi(b, \bar{a})$, for some $\varphi(x, \bar{y})$ and $\bar{a} \in A^{<\omega}$.

In other words, $\mathrm{Scl}^{\mathfrak{M}}(A)$ is the collection of the least witnesses in $\mathfrak{M}$ for all true statements of the form $\exists x \varphi(x, \bar{a})$, for $\bar{a}$ in $A^{<\omega}$. If it is clear from the context what $\mathfrak{M}$ is, we will usually drop the superscript $\mathfrak{M}$ from $\mathrm{Scl}^{\mathfrak{M}}(A)$.

If T is one of the theories mentioned above, then in every model of T, $\mathbb{N} \subseteq \mathrm{Scl}(\varnothing)$. Each natural number n is the only witness to the existential statement $\exists x \, \forall y[\neg \exists z \, \mathrm{Succ}(z, y) \implies \mathrm{Succ}_n(y, x)]$, i.e., n is the only n-th successor of 0.

In what follows we will consider $\mathrm{Scl}(A)$ not just as a set, but as a structure whose relations are restrictions to $\mathrm{Scl}(A)$ of the relations of $\mathfrak{M}$.

PROPOSITION 10.5. Let $\mathfrak{M}$ be a model of T. Then the following holds for all a, b, and all $A \subseteq M$

(1) If $a \in \mathrm{Scl}(A)$, then $a \in \mathrm{Scl}(B)$ for a finite $B \subseteq A$.
(2) If $B \subseteq \mathrm{Scl}(A)$ and $a \in \mathrm{Scl}(B)$, then $a \in \mathrm{Scl}(A)$. In particular $\mathrm{Scl}(\mathrm{Scl}(A)) = \mathrm{Scl}(A)$.
(3) $\mathrm{Scl}(A) \prec \mathfrak{M}$.

PROOF. Part (1) follows directly from the definition of Skolem closure. To prove (2), suppose that $a \in \mathrm{Scl}(B)$, for some $B \subseteq A$. By (1), we can assume that B is finite. To simplify notation, let us assume that $B = \{b\}$. The general argument is similar.

Since a is in $\mathrm{Scl}(B)$, there is a formula $\varphi(x, y)$ such that $\mathfrak{M} \models \mu_\varphi(a, b)$ and since b is in $\mathrm{Scl}(A)$, there are a formula $\psi(x, \bar{y})$ and a tuple $\bar{a}$ in $A^{<\omega}$ such that $\mathfrak{M} \models \mu_\psi(b, \bar{a})$. Then

$$\mathfrak{M} \models \exists x \exists z[\mu_\varphi(x, z) \wedge \mu_\psi(z, \bar{a})].$$

Let $\theta(x, \bar{y})$ be $\exists z[\mu_\varphi(x, z) \wedge \mu_\psi(z, \bar{y})]$. Then $\mathfrak{M} \models \mu_\theta(a, \bar{a})$; hence $b \in \mathrm{Scl}(A)$.

Part (3) follows directly from (2) and the Tarski-Vaught test (Theorem 6.9). □

Proposition 10.5 has a useful corollary. For each $A \subseteq M$, let $\mathfrak{M}(A)$ be $(\mathfrak{M}, a)_{a \in A}$.

COROLLARY 10.6. Let $\mathfrak{M}$ be a model of T. Then for each $A \subseteq M$, $\mathrm{Scl}^{\mathfrak{M}}(A) = \mathrm{Scl}^{\mathfrak{M}(A)}(\varnothing)$.

EXERCISE 10.7. Prove Corollary 10.6.

THEOREM 10.8 (Ehrenfeucht Mostowski Theorem). *Let $\mathfrak{M}$ be an infinite model of T, and let $(I, <)$ be a linearly ordered set. Then $\mathfrak{M}$ has an elementary extension $\mathfrak{N}$ such that $I \subseteq N$ and every automorphism of $(I, <)$ extends to an automorphism of $\mathfrak{N}$.*

PROOF. Let $\mathfrak{M}$ and $(I, <)$ as in the statement of the theorem. Let $T^* = \mathrm{Th}(M, a)_{a \in M}$. By Theorem 10.3, T^* has a model $\mathfrak{K}$ with a sequence of indiscernibles $A = \{a_i : i \in I\}$ that is order isomorphic to $(I, <)$. Since $(A, <)$ is isomorphic to $(I, <)$, we can assume that $I \subseteq K$.

Let $\mathfrak{N}$ be the Skolem closure of I in $\mathfrak{K}$. Notice that since $\mathfrak{K} \models T^*$, $\mathfrak{M} \prec \mathfrak{N}$. By Proposition 10.5, we also have that $\mathfrak{N} \prec \mathfrak{K}$.

Let α be an order preserving permutation of I. We extend α to $\beta : N \longrightarrow N$ as follows. For each $b \in N$ there are a formula φ of $\mathcal{L}_{T^*}$ and $\bar{a}$ in $I^{<\omega}$ such that $\mathfrak{N} \models \mu_\varphi(b, \bar{a})$. Because α is order preserving, I is a sequence of indiscernibles, and $\mathfrak{N} \models \exists! x \mu_\varphi(x, \bar{a})$, it follows that $\mathfrak{N} \models \exists! x \mu_\varphi(x, \alpha(\bar{a}))$. Let $\beta(b)$ be the unique b' such that $\mathfrak{N} \models \mu_\varphi(b', \alpha(\bar{a}))$.

Let let us check that β is well defined. If

$$\mathfrak{N} \models \mu_{\varphi_1}(b, \bar{a}_1) \wedge \mu_{\varphi_2}(b, \bar{a}_2),$$

then

$$\mathfrak{N} \models \forall x_1, x_2[\mu_{\varphi_1}(x_1, \bar{a}_1) \wedge \mu_{\varphi_2}(x_2, \bar{a}_2)] \Longrightarrow x_1 = x_2.$$

By indiscernibility of I,

$$\mathfrak{N} \models \forall x_1, x_2[\mu_{\varphi_1}(x_1, \alpha(\bar{a}_1)) \wedge \mu_{\varphi_2}(x_2, \alpha(\bar{a}_2))] \Longrightarrow x_1 = x_2.$$

This shows that $\beta(b)$ does not depend on the choice of the defining formula $\varphi(x, \bar{y})$ or the parameters in I.

Now let us verify that β is an automorphism. For more advanced readers, it is an exercise. For beginners, the details follow.

To simplify notation, let us consider a special case. Suppose $\mathfrak{N} \models \varphi(b_1, b_2)$. Then for some φ_1, φ_2 and $\bar{a}_1, \bar{a}_2$ in $I^{<\omega}$

$$\mathfrak{N} \models \mu_{\varphi_1}(b_1, \bar{a}_1) \wedge \mu_{\varphi_2}(b_2, \bar{a}_2).$$

Hence,

$$\mathfrak{N} \models \exists! x_1 \exists! x_2[\mu_{\varphi_1}(x_1, \bar{a}_1) \wedge \mu_{\varphi_2}(x_2, \bar{a}_2) \wedge \varphi(x_1, x_2)].$$

It follows that $\mathfrak{N} \models \varphi(\beta(b_1), \beta(b_2))$. Applying the same argument to $\neg\varphi(x_1, x_2)$ we get

$$N \models \varphi(b_1, b_2) \iff \varphi(\beta(b_1), \beta(b_2)).$$

In particular, this shows that β is one-to-one.

To prove that β is onto, consider a b in $\mathrm{Scl}(I)$ and φ and $\bar{a}$ in $I^{<\omega}$ such that $\mathfrak{N} \models \mu_\varphi(b, \bar{a})$. Then there is a c such that $\mathfrak{N} \models \mu_\varphi(c, \alpha^{-1}(\bar{a}))$. Hence, $\beta(c) = b$. $\qquad\square$

CHAPTER 11

Skolemization

Up to this point, we have discussed relational structures. Some arguments have involved functions, but always with the understanding that in the relational setting, functions must be represented by their graphs. For more advanced applications, this approach becomes tedious, so it is time to include function symbols in the formalism.

11.1. Functions

From now on, a structure will mean a domain M with sets of relations, constants, and functions $f : M^n \longrightarrow M$, for any $n > 0$. For each function in a structure we introduce a function symbol with assigned arity. All function symbols are interpreted as *total functions*, i.e., if f is an n-ary function symbol, then its interpretation in a structure $\mathfrak{M}$ is a function $f^{\mathfrak{M}} : M^n \longrightarrow M$. Most of the time, we will not distinguish between formal symbols and the informal names of their interpretations in structures.

In the relational setup, the language of fields includes two ternary relation symbols $+$ and $\cdot$. Instead, we will now consider them as binary functions. Later, we will keep the same symbols, but for a moment let us introduce two binary function symbols s and p. In a given field $\mathfrak{F}$, we interpret s and p as functions $s^{\mathfrak{F}} : F^2 \longrightarrow F$ and $p^{\mathfrak{F}} : F^2 \longrightarrow F$, defined by $s^{\mathfrak{F}}(a, b) = a + b$ and $p^{\mathfrak{F}}(a, b) = a \cdot b$. So far, not much has been accomplished. We just replaced relations by functions, but the point is that we have gained syntactic expressive power that will allow us to formalize algebraic constructions and arguments in a natural way.

Functions can be composed and that feature will now be introduced into the first-order formalism, by the following definition.

DEFINITION 11.1. Let $\mathcal{L}$ be a language with function symbols $f_1, \ldots, f_n$, where f_i has arity m_i. A sequence of symbols of $\mathcal{L}$ is a *term* if it is either a variable, or a constant symbol, or is of the form $f_i(t_1, \ldots, t_{m_i})$, where $t_1, \ldots, t_{m_i}$ are terms.

Terms are built inductively by composition, starting from the variables, constant symbols, and function symbols of the language.

For the field $\mathfrak{F}$ above, the term $s(p(x, y), p(x, x))$ represents the polynomial function $f(x, y) = xy + x^2$.

If we expand the language of fields by a function symbol of subtraction, then the terms of the language can be identified with the polynomials with natural number coefficients.

All polynomial functions can also be defined in the relational setting as well, but the definitions are much less natural. The reader should try the following exercise.

EXERCISE 11.2. Write a first-order definition of $f(x, y) = xy + x^2$, in the language with relation symbols $A(x, y, z)$ and $M(x, y, z)$ for addition and multiplication as relations.

If f and g are functions, then the statement $f(g(x)) = y$ can be formally written as

$$\exists z[g(x) = z \land f(z) = y]$$

and in this form it can be translated into a relational language. There is a price to pay though. The price is the quantifier $\exists$ that is absent in the original statement. The translation increases the number of quantifiers in formulas. The quantifier complexity of a formula is measured by the number of quantifiers in it. The larger the quantifier complexity of a formula, the more complex is the set defined by it. This last statement is not precise. First-order logic allows us to add any number of quantifiers to any formula, so in particular we can add dummy quantifiers to a formula without changing the sets the formula defines in structures. So more precisely: for many theories, for each $n > 0$, each model has subsets that can be defined by a formula with n quantifiers that cannot be defined with fewer than n quantifiers. Such are for example $\text{Th}((\mathbb{N}, +, \cdot))$ and Zermelo-Fraenkel set theory. For other, such as $\text{Th}((\mathbb{R}, +, \cdot))$ and $\text{Th}((\mathbb{C}, +, \cdot))$, it can be shown that every parametrically definable subset of the domain can be defined by a formula without quantifiers. In Chapter 8, we have proved that this is the case for $\text{Th}((\mathbb{N}, <))$.

Functional notation helps to reduce quantifier complexity of formulas, because some of that complexity is absorbed by the terms of the language that become parts of atomic formulas. For that, we need to modify the definition of atomic formula.

DEFINITION 11.3. If $\mathcal{L}$ is a language with function symbols, then the *atomic formulas* of $\mathcal{L}$ are all formulas of the form $t = s$ and $R(t_1, \ldots, t_n)$, where $t, s, t_1 \ldots, t_n$ are terms of $\mathcal{L}$ and R is an n-ary relation symbol of $\mathcal{L}$.

With function symbols included, there are more first-order formulas and Tarski's definition of truth must be extended to include them all. This is done in a natural way. A formal definition proceeds by induction. As before, induction begins with the atomic formulas, but prior to that evaluation of terms must be defined and this is done by induction on the complexity of terms.

Let α be an evaluation of the variables in the domain of a structure $\mathfrak{M}$. We will define $\alpha(t)$ for all terms t. Let t be a term. If $t = x_i$ then the evaluation of t is given by α. If t is a constant symbol c, then $\alpha(t) = c^{\mathfrak{M}}$. If t is neither a variable nor a constant symbol, then it is of the form $f(t_1, \ldots, t_n)$, where f_i is a function symbol and $t_1, \ldots, t_n$ are terms. If α is already defined for the terms of t and $\alpha(t_1) = a_1, \ldots, \alpha(t_n) = a_n$, then $\alpha(t) = f^{\mathfrak{M}}(a_1, \ldots, a_n)$.

Now, if t and s are terms, then $t = s$ is satisfied in $\mathfrak{M}$ by α if the evaluations of t and s are the same in M when each variable x_i that occurs either in t or in s is replaced by $\alpha(x_i)$; and similarly for formulas of the form $R(t_1, \ldots, t_n)$.

All model-theoretic results we have discussed so far hold for languages with function symbols as well. If proofs by induction on the complexity of formulas are involved, as in the proof of the compactness theorem, one can either add inductive steps to the atomic case, or just translate everything into a relational language. The results stay the same, but we need to modify the definition of substructure.

If $\mathfrak{M}$ and $\mathfrak{N}$ are structures for the same language $\mathcal{L}$ and $M \subseteq N$, then $\mathfrak{M}$ is a *substructure* of $\mathfrak{N}$ if, as before, for each n-ary relation symbol R in $\mathcal{L}$, $R^{\mathfrak{M}} = R^{\mathfrak{N}} \cap M^n$, for each constant symbol c, $c^{\mathfrak{M}} = c^{\mathfrak{N}}$ and, in addition, for each n-ary function symbol f and all $\bar{a}$ in M^n, $f^{\mathfrak{N}}(\bar{a}) = f^{\mathfrak{M}}(\bar{a})$. In particular, this means that for all $\bar{a}$ in M^n, $f^{\mathfrak{N}}(\bar{a})$ must be in M.

An important difference between model theories of relational and functional structures is illustrated by the following two exercises.

EXERCISE 11.4. Let $\mathfrak{F} = (F, +, \cdot, 0^{\mathfrak{F}}, 1^{\mathfrak{F}})$ be a field, where $+$ and $\cdot$ are binary functions, and let $\mathfrak{H}$ be a substructure of $\mathfrak{F}$. Show that $\mathfrak{H}$ is a field.

EXERCISE 11.5. Let $\mathfrak{Z} = (\mathbb{Z}, +, 0)$ be the group of integers, where $+$ is a ternary relation. Find a substructure of $\mathfrak{Z}$ that is not a group.

11.2. Built-in Skolem functions

The proof of the important Theorem 10.8 in the previous chapter was based on a construction involving indiscernible sequences and their Skolem closures. What mattered was that the theories T that we discussed, such as the complete theories of expansions of $(\mathbb{N}, <)$, have what is called *built-in Skolem functions*, meaning that for each formula $\varphi(x, \bar{y})$, there is a formula $\mu_\varphi(x, \bar{y})$, with free variables as shown, which in every model of T defines a witness function for the formula $\varphi(x, \bar{y})$. The formulas $\mu_\varphi(x, \bar{y})$ do not have to be defined as in the previous chapter, but must be such that for each formula $\varphi(x, \bar{y})$ and each model $\mathfrak{M}$ of T,

$$\mathfrak{M} \models \forall \bar{y}[\exists x \varphi(x, \bar{y}) \implies (\exists! x \mu_\varphi(x, \bar{y}) \wedge \exists x(\mu_\varphi(x, \bar{y}) \wedge \varphi(x, \bar{y})))]. \qquad (*)$$

If T is the complete theory of an expansion of $(\mathbb{N}, <)$, then every model of T satisfies the *least number principle*: for each formula $\varphi(x, \bar{y})$ of the language of T,

$$\forall \bar{y}[\exists x \varphi(x, \bar{y}) \implies \exists x(\varphi(x, \bar{y}) \wedge \forall z(z < x \implies \neg \varphi(z, \bar{y})))].$$

In other words, in every model of T, each nonempty definable set has a least element. The function that assigns to $\bar{y}$ the least element of the set defined by

$\varphi(x, \bar{y})$ is definable and we have taken advantage of it while proving the main results in the previous chapter.

Now we will fully switch to function notation. For any T that satisfies $(*)$ above, we define

$$f_\varphi(\bar{y}) = z \iff [(\exists x \varphi(x, \bar{y}) \wedge \mu_\varphi(z, \bar{y})) \vee (\forall x \neg \varphi(x, \bar{y}) \wedge \mu_{(x=x)}(z))].$$

We will call $f_\varphi(\bar{y})$ a *Skolem function* for $\varphi(x, \bar{y})$. The second disjunct in the definition is there to make sure that $f_\varphi(\bar{y})$ is a total function in each model of T.

In theories with built-in Skolem functions, $f_\varphi(\bar{y})$ is defined for all formulas $\varphi(x, \bar{y})$ of the language and that includes all formulas in which $\bar{y}$ is the empty sequence. In such a case f_φ is constant and its value in each model of T is either an element of the set defined by $\varphi(x)$ if this set in nonempty, or otherwise it is just some fixed element of the domain of the model.

It T has built-in Skolem function, then the functions $f_\varphi(x, \bar{y})$ may be defined in different ways and it is not important for applications which definitions we choose. We will examine one example below.

Let T be $\mathrm{Th}((\mathbb{N}, <))$, and let $\mathcal{L}$ be the language of T, i.e., $\{<\}$. For each formula $\varphi(x, \bar{y})$ and each $\bar{n}$ in $\mathbb{N}^{<\omega}$, let $X_{\varphi, \bar{n}}$ be the set defined by $\varphi(x, \bar{n})$ in $(\mathbb{N}, <)$. Let

$$f_\varphi(\bar{n}) = \begin{cases} \min X_{\varphi, \bar{n}}, & \text{if } X_{\varphi, \bar{n}} \text{ is nonempty} \\ 0, & \text{otherwise.} \end{cases}$$

and

$$g_\varphi(\bar{n}) = \begin{cases} \min X_{\varphi, \bar{n}}, & \text{if } X_{\varphi, \bar{n}} \text{ is infinite} \\ \max X_{\varphi, \bar{n}}, & \text{if } X_{\varphi, \bar{n}} \text{ is nonempty and finite} \\ 0, & \text{otherwise.} \end{cases}$$

We could have defined $f_\varphi(\bar{y})$ and $g_\varphi(\bar{y})$ for all models of T, but here we just want to compare the following two expansions of $(\mathbb{N}, <)$. Let $\mathfrak{N}_f = (\mathbb{N}, <, f_\varphi)_{\varphi \in \mathcal{L}}$ and $\mathfrak{N}_g = (\mathbb{N}, <, g_\varphi)_{\varphi \in \mathcal{L}}$.

Both $\mathfrak{N}_f$ and $\mathfrak{N}_g$ are what is called *definitional* expansions, because all added functions are definable in $(\mathbb{N}, <)$. For each n, if a subset of $\mathbb{N}^n$ is parametrically definable in any of these expansions, then it is already parametrically definable in $(\mathbb{N}, <)$. From this point of view we did not change the original structure much; we only made some definable functions directly visible. Still, something has changed: $\mathfrak{N}_f$ and $\mathfrak{N}_g$ are not elementarily equivalent.

EXERCISE 11.6. Prove the last statement above.

DEFINITION 11.7. Let $\mathfrak{M}$ be a model of a theory T with built-in Skolem functions. For each $A \subseteq M$, the *Skolem closure* of A in $\mathfrak{M}$, denoted $\mathrm{Scl}^{\mathfrak{M}}(A)$, is the set $\{f_\varphi(\bar{a}) : \varphi \in \mathcal{L}_T \text{ and } \bar{a} \in A^{<\omega}\}$.

As before, we will drop the superscript $\mathfrak{M}$ if it is clear from the context and we will consider $\mathrm{Scl}(A)$ as a structure whose relations are restrictions to $\mathrm{Scl}(A)$ of relations of $\mathfrak{M}$.

The proofs of the next two results follow the proofs of Proposition 10.5 and Theorem 10.8. Parts (4) and (5) of Proposition 11.8 is new and its proof is left for an exercise.

PROPOSITION 11.8. Let $\mathfrak{M}$ be a model of a theory with built-in Skolem functions. Then the following holds for all a, b and all $A \subseteq M$

(1) If $a \in \mathrm{Scl}(A)$, then $a \in \mathrm{Scl}(B)$ for a finite $B \subseteq A$.
(2) If $B \subseteq \mathrm{Scl}(A)$ and $a \in \mathrm{Scl}(B)$, then $a \in \mathrm{Scl}(A)$. In particular $\mathrm{Scl}(\mathrm{Scl}(A)) = \mathrm{Scl}(A)$.
(3) $\mathrm{Scl}(A) \prec \mathfrak{M}$.
(4) Let $\mathfrak{K}$ be a substructure of $\mathfrak{M}$, then $\mathfrak{K} \prec \mathfrak{M}$ if and only if K is closed under Skolem functions, for all n-ary Skolem functions $f_\varphi(\bar{y})$ and all $\bar{a}$ in K^n, for $n > 0$, $f_\varphi(\bar{a})$ is in K.
(5) If $\mathfrak{K} \prec \mathfrak{M}$, then for all $A \subseteq K$, $\mathrm{Scl}^{\mathfrak{K}}(A) = \mathrm{Scl}^{\mathfrak{M}}(A)$.

THEOREM 11.9. *Let $\mathfrak{M}$ be a model with an infinite domain of a theory with built-in Skolem functions, and let $(I, <)$ be a linearly ordered set. Then $\mathfrak{M}$ has an elementary extension $\mathfrak{N}$ such that $I \subseteq M$ and every automorphism of $(I, <)$ extends to an automorphism of $\mathfrak{N}$.*

EXERCISE 11.10. Prove Proposition 11.8 and Theorem 11.9.

11.2.1. Skolem closures and definable elements. Notice that if the set defined by $\varphi(x, \bar{a})$ in a model $\mathfrak{M}$ is nonempty, then $f_\varphi(\bar{a})$ is definable in $(\mathfrak{M}, \bar{a})$. In particular, for each parameter-free formula $\varphi(x)$ defining a nonempty set, f_φ is a definable element of M. The converse is also true. If a formula $\varphi(x, \bar{a})$ defines b in $(\mathfrak{M}, \bar{a})$ for some $\bar{a}$ in $M^{<\omega}$, then $f_\varphi(\bar{a}) = b$; hence b is in $\mathrm{Scl}^{\mathfrak{M}}(\bar{a})$. These remarks prove the following proposition.

PROPOSITION 11.11. If $\mathfrak{M}$ is a model of a theory T with built-in Skolem functions, then for each $A \subseteq M$, $\mathrm{Scl}(A)$ is the set of elements that are definable in $(\mathfrak{M}, a)_{a \in A}$.

By Proposition 11.8 (3), this gives us the following corollary.

COROLLARY 11.12. Every theory with built-in Skolem functions has a model all of whose elements are definable.

PROPOSITION 11.13. Let $\mathfrak{M}$ be a model of a theory T with built-in Skolem functions. If M is infinite, then $\mathrm{Scl}(\varnothing)$ is infinite. If M is finite, then $\mathrm{Scl}(\varnothing) = M$.

PROOF. The proposition also follows from Proposition 11.8(3), but we will give a direct proof. Let $\varphi_0(x)$ be $x = x$, and let $a_0 = f_{\varphi_0}$. Then, for each n, let $\varphi_{n+1}(x)$ be $\bigwedge\{\neg(x = a_i) : i \leq n\}$. If for each $i \leq n$, a_i is definable in $\mathfrak{M}$, then so is a_{n+1}. By induction, it follows that for each n, a_n is definable in $\mathfrak{M}$. Hence, $\{a_n : n \in \mathbb{N}\} \subseteq \mathrm{Scl}(\varnothing)$.

If M is finite, then there is an n such that $M = \{a_0, \dots, a_n\}$. If M is infinite, then for each n, $a_{n+1} \notin \{a_0, \dots, a_n\}$. $\qquad\square$

EXERCISE 11.14. Let T be a complete theory in a countable language that has built-in Skolem functions. Prove that if T has a model with an infinite domain, then T is not $\aleph_0$-categorical.

11.2.2. Lattices of elementary substructures. Proposition 11.8 (4) makes parts of model theory of theories with built-in Skolem functions similar to algebra. In algebra, we study subalgebras (subgroups, subfields, and such) that are generated by applying algebraic operations to elements of subsets of a given algebra. In model theory, we study elementary substructures that are generated by applying Skolem functions to elements of subsets of the domain of a given structure.

For the rest of this section let T be a theory with built-in Skolem functions, and let $\mathfrak{M}$ be a model of T.

Let $\mathrm{Lt}(\mathfrak{M})$ be $(\{\mathfrak{K} : \mathfrak{K} \prec \mathfrak{M}\}, \preceq)$, where $\mathfrak{K} \preceq \mathfrak{M}$ means that either $\mathfrak{K} \prec \mathfrak{M}$ or $\mathfrak{K} = \mathfrak{M}$.

LEMMA 11.15. *Let $\mathfrak{K}$ and $\mathfrak{L}$ be elementary submodels of $\mathfrak{M}$. Then $\mathrm{Scl}(K \cap L) = K \cap L$.*

PROOF. Let $\bar{a}$ be an n-tuple in $(K \cap L)^n$, and let $f_\varphi(\bar{y})$ be an n-ary Skolem term. Because $\mathfrak{K}$ and $\mathfrak{L}$ are elementary submodels of $\mathfrak{M}$, by Proposition 11.8 (4), $f_\varphi(\bar{a})$ is in K and it is L, so $f_\varphi(\bar{a})$ is in $K \cap L$. $\qquad\square$

Lemma 11.15 shows that $K \cap L$ is the domain of an elmentary submodel of $\mathfrak{M}$. We will refer to this model as $\mathfrak{K} \cap \mathfrak{L}$. We have an immediate corollary.

COROLLARY 11.16. *$\mathfrak{K} \cap \mathfrak{L}$ is the largest elementary submodel of both $\mathfrak{K}$ and $\mathfrak{L}$.*

A *partial ordering* is a binary relation $\preceq$ on a set A that satisfies the following conditions for all a, b, and c in A:

(1) $a \preceq a$ ($\preceq$ is reflexive);
(2) if $a \preceq b$ and $b \preceq a$, then $a = b$ ($\preceq$ is antisymmetric);
(3) if $a \preceq b$ and $b \preceq c$, then $a \preceq c$ ($\preceq$ is transitive).

A partial ordering $(A, \preceq)$ is a *lattice* if for each pair a, b in L, there are c and d in L such $a \preceq c$, $b \preceq c$, and c is the smallest such element in A; and $d \preceq a$ and $d \preceq b$, and d is the largest such element in A.

Let $\mathfrak{M}$ be a model of T. Let $\mathfrak{K}$ be an elementary submodel of $\mathfrak{M}$. If $K = \mathrm{Scl}^{\mathfrak{M}}(A)$, for some $A \subseteq M$, then we say that K is *generated* by A in $\mathfrak{M}$.

PROPOSITION 11.17. $\mathrm{Lt}(\mathfrak{M})$ is a lattice.

PROOF. Let $\mathfrak{K}$ and $\mathfrak{L}$ be elementary submodels of $\mathfrak{M}$. Then, by Corollary 11.16, $\mathfrak{K} \cap K$ is the largest model in $\mathrm{Lt}(\mathfrak{M})$ that is smaller than $\mathfrak{K}$ and $\mathfrak{L}$ and $\mathrm{Scl}^{\mathfrak{M}}(K \cup L)$ is the smallest model in $\mathrm{Lt}(\mathfrak{M})$ that extends both $\mathfrak{K}$ and $\mathfrak{L}$. $\square$

For a theory T, one can ask a general question: What lattices can be represented as $\mathrm{Lt}(\mathfrak{M})$, where $\mathfrak{M} \models T$? This question has been studied for nonstandard models of arithmetic, with very interesting results, most of them summarized in [KS06, Chapter 4].

We finish this section with an exercise that illustrates what is different for theories without built-in Skolem functions. One such theory is DLO.

EXERCISE 11.18. Let D be a dense subset of $\mathbb{Q}$. Show that $(D, <) \prec (\mathbb{Q}, <)$. Find two dense disjoint subsets of $\mathbb{Q}$. Compare this example with Corollary 11.16.

11.3. Skolemizing Theories and Structures

In this section we will show how to prove Theorem 11.9 for theories without built-in Skolem functions. This is done by brute force. If a theory does not have built-it Skolem functions, we will just add new function symbols to the language and expand the theory by adding appropriate axioms that will make the functions interpreting the new symbols Skolem functions. The difference is that the new functions will be Skolem functions only for formulas of the original language of the theory.

DEFINITION 11.19. For a theory T in a language $\mathcal{L}$, let *Skolemization* of T, denoted $\mathrm{Sk}(T)$, be the theory in the language $\mathcal{L}$ with additional function symbols $f_\varphi(\bar{y})$, one for each formula $\varphi(x, \bar{y})$ of $\mathcal{L}$, obtained by adding to the axioms of T all sentences of the form

$$\forall \bar{y}[\exists x \varphi(x, \bar{y}) \implies \varphi(f_\varphi(\bar{y}), \bar{y})].$$

The extended language will be denoted by $\mathrm{Sk}(\mathcal{L})$.

If T has built-in Skolem functions, then we can use those functions in every model of T and there is no need to expand the language. If T does not have such functions, still any model $\mathfrak{M}$ of T can be expanded to a model $\mathrm{Sk}(\mathfrak{M})$ by letting each $f_\varphi(\bar{a})$ be an element of the set defined by $\varphi(x, \bar{a})$ in $\mathfrak{M}$, for all $\bar{a}$ for which this set is nonempty and letting $f_\varphi(\bar{a})$ be some fixed element c of M otherwise. We call $\mathrm{Sk}(\mathfrak{M})$ a *Skolemization* of $\mathfrak{M}$. In both cases, we call the interpretations of the symbols $f_\varphi(\bar{y})$ in $\mathfrak{M}$, *Skolem functions* of $\mathfrak{M}$.

It is time now for a set-theoretic digression. Each structure either has built-in Skolem functions, or can be expanded to a structure that has Skolem functions for the formulas of the original language, but there is a technical problem of a set-theoretic nature that needs to be addressed.

To define Skolem functions in a structure $\mathfrak{M}$, we need to select one element from each nonempty parametrically definable subset of M. If the set defined by $\varphi(x, \bar{a})$ is nonempty, then we have said that $f_\varphi(\bar{a})$ can be any element of this set. This is fine, but how do we know that a function that assigns such an element to every $\bar{a}$ exists? The statement that such a function exists resembles the axiom of choice, which, in one of many equivalent formulations, says that for any sequence of nonempty sets $\{X_i\}_{i \in I}$, there is a function $f : I \longrightarrow \bigcup_{i \in I} X_i$, such that for each i, $f(i) \in X_i$. Such a function f is called a choice function for $\{X_i\}_{i \in I}$. If we assume the axiom of choice, we get our Skolem functions, one for each sequence of nonempty sets defined by $\varphi(x, \bar{a})$, indexed by the $\bar{a}$ in $M^{<\omega}$ of appropriate length.

To prove that every structure can be Skolemized, we can appeal to the axiom of choice. It can be shown that it is also necessary. If one assumes that every structure can be Skolemized, then one can prove the axiom of choice from the other axioms of ZF. Below is an outline of the argument, for a full proof one has to check that the axioms of ZF suffice to carry out the argument.

Let $\{X_i\}_{i \in I}$ be a sequence of nonempty sets. Let $X = \bigcup_{i \in I} X_i$, and let $(X \cup I, R)$ be the structure with the domain $X \cup I$ and one binary relation R defined by $(x, i) \in R$ if and only if $x \in X_i$. Let $\varphi(x, y)$ be the atomic formula $R(x, y)$. Then, for every i in I, $f_{\varphi(x,y)}(i) \in X_i$, so $f_{\varphi(x,y)}(y)$ restricted to I is a choice function for $\{X_i\}_{i \in I}$.

Because every automorphism of an expansion of a structure is also an automorphism of the structure, every automorphism of $\mathrm{Sk}(\mathfrak{M})$ is also an automorphism of $\mathfrak{M}$. Hence, as an immediate corollary of Theorem 11.9 we get the following theorem.

THEOREM 11.20. *Let $\mathfrak{M}$ be a model with an infinite domain, and let $(I, <)$ be a linearly ordered set. Then $\mathfrak{M}$ has an elementary extension $\mathfrak{N}$ such that every automorphism of $(I, <)$ extends to an automorphism of $\mathfrak{N}$.*

PROOF. Apply Theorem 11.9 to $\mathrm{Sk}(\mathfrak{M})$.   □

All definable sets in a Skolemized model $\mathrm{Sk}(\mathfrak{M})$ for a theory with built-in Skolem function are already defined in $\mathfrak{M}$, $\mathrm{Aut}(\mathrm{Sk}(\mathfrak{M})) = \mathrm{Aut}(\mathfrak{M})$. For theories without built-in Skolem functions, the situation is dramatically different. Let $\mathfrak{M} = (M)$, be the structure with an infinite domain M, with no functions, relations, nor constants. Then, $\mathrm{Aut}(\mathfrak{M})$ is the group of all permutations of M. By Propositions 11.11 and 11.13, $X = \mathrm{Scl}^{\mathrm{Sk}(\mathfrak{M})}(\varnothing)$ is infinite and each element of X is fixed by every automorphismof $\mathrm{Sk}(\mathfrak{M})$. If M is countable then there is a Skolemization $\mathrm{Sk}(\mathfrak{M})$ such that $X = M$. In this case $\mathfrak{M}$ is rigid.

EXERCISE 11.21. Let M be a countable set, and let $\mathfrak{M} = (M)$. Define $\mathrm{Sk}(\mathfrak{M})$ such that $\mathfrak{M}$ is rigid.

11.3.1. Löwenheim-Skolem Theorem. Skolemization can be used for a quick proof of the Löwenheim-Skolem theorem.

Let $\mathfrak{M}$ be a structure for a language $\mathcal{L}$, and let A be a subset of M. We define the Skolem closure of A, relative to a Skolemization $\mathrm{Sk}(\mathfrak{M})$, as the smallest subset of M that is closed under all Skolem functions. Then, as in Proposition 11.8, it can be shown that $\mathrm{Scl}(A)$, considered as an $\mathcal{L}$-substructure of $\mathfrak{M}$, is an elementary substructure of $\mathfrak{M}$. As an immediate corollary, we get the following version of the Löwenheim-Skolem theorem.

THEOREM 11.22. *Let $\mathfrak{M}$ be a structure for a countable language $\mathcal{L}$. If $A \subseteq M$ is countable, then there is a countable $\mathfrak{K}$ such that $A \subseteq K$ and $\mathfrak{K} \prec \mathfrak{M}$.*

PROOF. Let $\mathrm{Scl}(A)$ be the Skolem closure of A in some Skolemization of $\mathfrak{M}$. Since $\mathcal{L}$ is countable, $|\mathrm{Scl}(A)| = \aleph_0$, the result follows from the remark preceding the theorem. $\square$

Theorem 11.22 is formulated for countable languages and countable sets $A \subseteq M$. With a bit of set theory it can be generalized to arbitrary $\mathcal{L}$ and A, with the conclusion that $|K| \leq |A| + |\mathcal{L}|$.

Theorem 11.22 is called the downward Löwenheim-Skolem theorem, as it guarantees existence of smaller elementary submodels in a larger one. There is also the upward Löwenheim-Skolem theorem that goes in the other direction and it is an easy consequence of the compactness theorem for uncountable languages and the downward Löwenheim-Skolem theorem.

THEOREM 11.23. *If M is infinite, then for every cardinal $\kappa \geq |M|$, $\mathfrak{M}$ has an elementary extension $\mathfrak{N}$ such that $|N| = \kappa$.*

CHAPTER 12

Arithmetic

In previous chapters, examples of simple structures have been used to introduce basic model-theoretic techniques and to explain how they are used to study definability. This chapter is about nonstandard models of the theory of one familiar, but very complex structure, the *standard model of arithmetic* $(\mathbb{N}, +, \cdot)$.

12.1. True Arithmetic

Let TA be the complete theory of the standard model. TA stands for *true arithmetic*. *Peano arithmetic*, abbreviated PA, is a first-order theory in the language of the standard model that consists of a finite set of basic axioms for addition and multiplication, and the induction schema: for each formula $\varphi(x, \bar{y})$

$$\forall \bar{y}[(\varphi(0, \bar{y}) \wedge \forall x(\varphi(x, \bar{y}) \implies \varphi(x + 1, \bar{y}))) \implies \forall x \varphi(x, \bar{y})].$$

The basic axioms can be formulated in many equivalent ways. See [**Kay91**] for a full discussion.

All axioms of PA hold in the standard model, hence PA $\subseteq$ TA. By Gödel's incompleteness theorem, PA is a proper subset of TA.

All results about models of TA that we will discuss in this chapter hold for models of PA, but the proofs require checking that the facts about the standard model that we will use can be formalized as theorems of PA.

We say that a set of natural numbers is *arithmetic* if it is definable in the standard model. A function $f : \mathbb{N}^n \longrightarrow \mathbb{N}$ is arithmetic if its graph is an arithmetic subset of $\mathbb{N}^{n+1}$. In Chapter 4 we have seen that the subsets of $\mathbb{N}$ that are definable in $(\mathbb{N}, +)$ are exactly the ultimately periodic sets and that sets definable in $(\mathbb{N}, \cdot)$ are invariant under permutations of primes so that even the ordering of the natural numbers is not definable. When we combine addition and multiplication, the expressive power of the resulting language increases dramatically and that is because each natural number codes information that can be recovered with the aid of addition and multiplication. In particular, it can be shown that all computable functions are definable, but this is just a beginning. All computable functions are definable by formulas of low quantifier complexity. Arithmetic functions and sets form a hierarchy with respect to the number of quantifiers in their definitions. For each n, there are sets that can be defined by a formula with $n + 1$ quantifiers that cannot be defined by any formula with only n quantifiers. Richard Kaye's book [**Kay91**] gives a detailed presentation of all these results.

In this chapter we will go over details of Skolem's construction of an elementary extension of the standard model, and then we will use that extension to give a model-theoretic proof of Tarski's celebrated theorem on *undefinability of truth*, which in its simplest version says that the set of codes of sentences that are true in the the the standard model is not arithmetic, but first, to formulate the theorem, we need to introduce arithmetic coding.

12.2. Coding

In set theory one can prove that every infinite set X is in one-to-one correspondence with its Cartesian square X^2. Georg Cantor noticed that for $\mathbb{N}$ and $\mathbb{N}^2$ there is such a correspondence of a particularly simple form. It is given by Cantor's pairing function, which to every ordered pair of numbers (x, y) assigns its code $\langle x, y \rangle$ given by

$$\langle x, y \rangle = \frac{1}{2}(x + y)(x + y + 1) + y.$$

The function $(x, y) \mapsto \langle x, y \rangle$ is arithmetic. Its graph is defined by

$$\langle x, y \rangle = z \iff 2z = (x + y)(x + y + 1) + y.$$

Using Cantor's function, for each $n > 0$, we can recursively define a one-to-one correspondence of $\mathbb{N}^n$ and $\mathbb{N}$ by

$$\langle x_1, \ldots, x_n, x_{n+1} \rangle = \langle \langle x_1, \ldots, x_n \rangle, x_{n+1} \rangle.$$

All those coding functions are arithmetic, hence, for each $n > 1$, every arithmetic subset of $\mathbb{N}^n$ can be coded by an arithmetic subset of $\mathbb{N}$.

EXERCISE 12.1. Write a proof of the last sentence above.

Cantor's coding does a lot, but even more can be done with coding devised by Gödel. Cantor's functions allow us to code any finite sequence of natural numbers by a single number, but for sequences of different lengths the coding protocol is different. In Gödel's coding all finite sequences are coded uniformly.

Let $\bar{k} = k_1, \ldots, k_n$ be a sequence of natural numbers. Let $b = \max\{n, k_1, \ldots, k_n\}$ and let $m = b!$. Then numbers $m + 1, 2m + 1, \ldots, nm + 1$ are relatively prime. By the Chinese reminder theorem, there is an a such that for all $i = 1, \ldots n$,

$$a \equiv k_i \mod (im + 1),$$

i.e., k_i is the reminder in division of a by $im + 1$. Finally let the code of $\bar{k}$ be $c = \langle a, m \rangle$.

From c we can recover the entire sequence $k_1, \ldots, k_n$. Moreover the relation "k is the i-th term of the sequence coded by c" is arithmetic. Let $(x)_y = z$ be an arithmetic formula expressing that z is the y-th term of the sequence coded by x.

While the number theory involved in Gödel's coding is elementary, the "moreover" in the last claim above should not be taken lightly. It is not an exercise for the reader. All details are patiently explained in [**Kay91**, Chapter 5].

Equipped with Gödel's coding, we can turn recursive definitions into first-order arithmetic definitions. A formal proof will not be given here, but let us see an example of how it works.

EXAMPLE 12.2. The exponential function is defined by recursion as follows: $2^0 = 1$ and $2^{n+1} = 2 \cdot 2^n$. The arithmetic formula below defines the relation $y = 2^x$ explicitly.

$$\exists z[(z)_0 = 1 \wedge \forall i(i < x \implies (z)_{i+1} = 2 \cdot (z)_i) \wedge (z)_x = y].$$

12.3. Definable Ultrapowers

An elementary extension $\mathfrak{M} = (M, +, \cdot)$ of the standard model is *simple* if it is a Skolem closure of a single element, i.e., there is a c in M such that $\mathfrak{M} = \mathrm{Scl}(c)$. A simple extension of the standard model can be obtained by first taking an elementary extension, and then taking the Skolem closure of any nonstandard element in it. Still, in this section, we will take a detour to construct a simple extension directly, by a construction due to Skolem.

Let $\{\varphi_n(x)\}_{n \in \mathbb{N}}$ be an enumeration of all formulas of the language of arithmetic with one free variable. For a formula $\varphi(x)$, $\varphi(\mathbb{N})$ will denote the set that is defined by $\varphi(x)$ in the standard model.

We will inductively define a nested sequence of infinite arithmetic sets $X_0 \supseteq X_1 \supseteq \cdots$. Let $X_0 = \mathbb{N}$, and suppose that $X_0, \ldots, X_n$ have been defined. Consider $\varphi_n(x)$. Since X_n is infinite, at least one of $X_n \cap \varphi_n(\mathbb{N})$ and $X_n \cap \neg\varphi_n(\mathbb{N})$ is infinite. We choose X_{n+1} to be one of those sets that is infinite.

Let $\mathrm{Def}(\mathbb{N})$ be the set of all arithmetic subsets of $\mathbb{N}$, and let

$$\mathcal{U} = \{D : D \in \mathrm{Def}(\mathbb{N}) \wedge \exists n(X_n \subseteq D)\}.$$

A straightforward proof of the next proposition is left as an exercise.

LEMMA 12.3. *Let X and Y be arithmetic sets, and let $\mathcal{U}$ be as defined above.*

(1) *If $X \in \mathcal{U}$ and $X \subseteq Y$, then $Y \in \mathcal{U}$.*
(2) *If $X \in U$ and $Y \in U$, then $X \cap Y \in \mathcal{U}$.*
(3) *Either $X \in \mathcal{U}$ or $(\mathbb{N} \setminus X) \in \mathcal{U}$.*

If $f : \mathbb{N} \longrightarrow \mathbb{N}$ and $g : \mathbb{N} \longrightarrow \mathbb{N}$ are arithmetic, then we will say that f and g are equivalent, written $f \sim g$, if the set $\{n : f(n) = g(n)\}$ is in $\mathcal{U}$. We will show that $\sim$ is an equivalence relation.

Since $\mathbb{N} \in \mathcal{U}$, we have that $f \sim f$; hence $\sim$ is reflexive.

If $f \sim g$, then clearly $g \sim f$; hence $\sim$ is symmetric.

From Lemma 12.3 it follows that if $f \sim g$ and $g \sim h$, then $f \sim h$; hence $\sim$ is transitive.

For each arithmetic function $f : \mathbb{N} \longrightarrow \mathbb{N}$, let $[f]$ be the equivalence class of f, i.e $\{g : f \sim g\}$.

Let M be the set of all equivalence classes $[f]$ for arithmetic f. The set M will be the domain of an elementary extension of the standard model. We need to define addition and multiplication on M. For $[f]$ and $[g]$ in M let

$$[f] + [g] = [h] \text{ iff } \{n : f(n) + g(n) = h(n)\} \in \mathcal{U},$$

and

$$[f] \cdot [g] = [h] \text{ iff } \{n : f(n) \cdot g(n) = h(n)\} \in \mathcal{U}.$$

Let $\mathfrak{M}$ be $(\mathfrak{M}, +, \cdot)$, with $+$ and $\cdot$ defined above. Using Lemma 12.3, one can show that $+$ and $\cdot$ are well-defined, i.e., they do not depend on the choice of representatives for each equivalence class.

For each $n \in \mathbb{N}$, let c_n be the constant function $c_n(x) = n$. It is easy to see that $m + n = k$ if and only if $[c_m] + [c_n] = [c_k]$ and similarly for multiplication. Hence, the function $n \mapsto [c_n]$ is an embedding of the standard model into $\mathfrak{M}$. We will identify the standard model with its image under this embedding. This way the standard model becomes a substructure of $\mathfrak{M}$.

The crucial property of $\mathfrak{M}$ is stated in the next lemma. It implies that $\mathfrak{M}$ is an elementary extension of the standard model.

LEMMA 12.4. *For every arithmetic formula $\varphi(x_1, \ldots, x_k)$ and all $[f_1], \ldots, [f_k]$ in $\mathfrak{M}$*

$$\mathfrak{M} \models \varphi([f_1], \ldots, [f_k]) \text{ iff } \{n : (\mathbb{N}, +, \cdot) \models \varphi(f_1(n)), \ldots, \varphi(f_k(n)))\} \in \mathcal{U}.$$

PROOF. The proof rests on Lemma 12.3. It proceeds by induction on the rank of φ, with the base case being the definition of $+$ and $\cdot$ in $\mathfrak{M}$.

As before, for simplicity of notation we will just consider the case of a single $[f]$ in $\mathfrak{M}$.

Suppose the lemma holds for ψ and θ, and $\varphi = \psi \wedge \theta$. Then

$$\mathfrak{M} \models \varphi([f]) \text{ iff}$$
$$\mathfrak{M} \models \psi([f]) \wedge \theta([f]) \text{ iff}$$
$$\mathfrak{M} \models \psi([f]) \text{ and } \mathfrak{M} \models \theta([f]) \text{ iff}$$
$$\{n : (\mathbb{N}, +, \cdot) \models \psi(f(n))\} \in \mathcal{U} \text{ and } \{n : (\mathbb{N}, +, \cdot) \models \theta(f(n))\} \in \mathcal{U} \text{ iff}$$
$$\{n : (\mathbb{N}, +, \cdot) \models \psi(f(n))\} \cap \{n : (\mathbb{N}, +, \cdot) \models \theta(f(n))\} \in \mathcal{U} \text{ iff}$$
$$\{n : (\mathbb{N}, +, \cdot) \models \psi(f(n)) \wedge \theta(f(n))\} \in \mathcal{U} \text{ iff}$$
$$\{n : (\mathbb{N}, +, \cdot) \models \varphi(f(n))\} \in \mathcal{U}.$$

If $\varphi = \neg\psi$, then

$$\mathfrak{M} \models \varphi([f]) \text{ iff}$$
$$\mathfrak{M} \not\models \psi([f]) \text{ iff}$$
$$\{n : (\mathbb{N}, +, \cdot) \models \psi(f(n))\} \notin \mathcal{U} \text{ iff}$$
$$\mathbb{N} \setminus \{n : (\mathbb{N}, +, \cdot) \models \psi(f(n))\} \in \mathcal{U} \text{ iff}$$
$$\{n : (\mathbb{N}, +, \cdot) \models \neg\psi(f(n))\} \in \mathcal{U} \text{ iff}$$
$$\{n : (\mathbb{N}, +, \cdot) \models \varphi(f(n))\} \in \mathcal{U}.$$

Finally, if $\varphi = \exists x\psi$, then

$$\mathfrak{M} \models \varphi([f]) \text{ iff}$$

$$\mathfrak{M} \models \exists x\psi(x, [f]) \text{ iff}$$

$$\mathfrak{M} \models \psi([g], [f]), \text{ for some } g, \text{ iff}$$

$$\{n : (\mathbb{N}, +, \cdot) \models \psi(g(n), f(n))\} \in \mathcal{U} \text{ for some } g, \text{ iff} \tag{4}$$

$$\{n : (\mathbb{N}, +, \cdot) \models \exists x\psi(x, f(n))\} \in \mathcal{U}. \tag{5}$$

The last two lines in the proof require an explanation. It is clear that (4) implies (5). So let us assume that

$$\{n : (\mathbb{N}, +, \cdot) \models \exists x\psi(x, f(n))\} = D \in \mathcal{U},$$

holds. Then we define $g : \mathbb{N} \longrightarrow \mathbb{N}$ by

$$g(n) = \begin{cases} \min\{x : \psi(x, f(n))\} & \text{if } n \in D, \\ 0, & \text{otherwise} \end{cases}$$

The function g is arithmetic and if $E = \{n : (\mathbb{N}, +, \cdot) \models \psi(g(n), f(n))\}$, then $D \subseteq E$, hence E is in $\mathcal{U}$, which gives the implication (5) $\Longrightarrow$ (4). $\qquad\square$

We did not define the ordering of $\mathfrak{M}$, but it is not necessary because $<$ is definable in the standard model, and, by elementarity of the extension, the same definition defines a linear ordering of $\mathfrak{M}$ such that for all $[f]$, $[g]$ in M, if $[f] < [g]$, then $[f] + [h] < [g] + [h]$ and if $[h] \neq [0]$, then $[f] \cdot [h] < [g] \cdot [h]$.

For the next two propositions, let i in M be the equivalence class of the identity function on $\mathbb{N}$.

PROPOSITION 12.5. For each standard m, $\mathfrak{M} \models m < [i]$.

PROOF. By Lemma 12.4,

$$\mathfrak{M} \models m < [i] \text{ iff } \{k : (\mathbb{N}, +, \cdot) \models m < k\} \in \mathcal{U}.$$

The formula $(m < x)$ is listed as $\varphi_n(x)$ in the enumeration of all arithmetic formulas used in the construction of the sequence $X_0 \supseteq X_1 \supseteq \cdots$ that we used to define $\mathcal{U}$. Because $\{k : (\mathbb{N}, +, \cdot) \models \neg(m < k)\}$ is finite, X_{n+1} is $X_n \cap \varphi_n(\mathbb{N})$ and this finishes the proof. $\qquad\square$

Proposition 12.5 shows that $\mathfrak{M}$ is a proper elementary extension of the standard model. We will use this $\mathfrak{M}$ in the proof of Tarski's theorem and for the proof we need to verify that $\mathfrak{M}$ is a simple extension.

PROPOSITION 12.6. $\mathrm{Scl}(i) = M$.

PROOF. Let a be an element of M. Then, $a = [f]$ for some arithmetic function f. Let $\varphi(x, y)$ be $x = f(y)$. By Lemma 12.4, $\mathfrak{M} \models \mu_\varphi([f], i)$ if and only if $\{n : (\mathbb{N}, +, \cdot) \models \mu_\varphi(f(n), n)\} \in \mathcal{U}$ (see Definition 10.4).

Because $\{n : (\mathbb{N}, +, \cdot) \models \mu_\varphi(f(n), n)\} = \mathbb{N}$ and $\mathbb{N}$ is in $\mathcal{U}$, we have $\mathfrak{M} \models \mu_\varphi([f], i)$; hence, $a \in \mathrm{Scl}(i)$. $\qquad\square$

We will use $\mathfrak{M}$ to prove an undefinability result about the standard model, but, unlike in the proof of minimality of $(\mathbb{N}, <)$, we cannot use automorphisms because $\mathfrak{M}$ is rigid. That $\mathfrak{M}$ is rigid follows from a result known as the Ehrenfeucht lemma [**Ehr73**]. The lemma implies that if $\mathfrak{M}$ is a simple elementary extension of the the standard model and $M = \mathrm{Scl}(c)$, then for all a, if $a \neq c$, then $\mathrm{tp}(a) \neq \mathrm{tp}(c)$. Therefore, if f is an automorphism of $\mathfrak{M}$, then $f(c) = c$, and since every a in $\mathrm{Scl}(c)$ is definable in $(\mathfrak{M}, c)$, we must have $f(a) = a$ as well.

Finally, a few words about the title of this section. A set $\mathcal{U}$ of subsets of $\mathbb{N}$ that satisfies the conditions in Lemma 12.3 for all sets X and Y, not just for the arithmetic ones, is called an *ultrafilter*. If $\mathcal{U}$ is an ultrafilter, then one can define the equivalence relation $\sim$ as above for all functions $f : \mathbb{N} \longrightarrow \mathbb{N}$ and the structure $\mathfrak{M}$ whose domain is the set of all equivalence classes $[f]$. Then, Lemma 12.4 holds with exactly the same proof. Such an $\mathfrak{M}$ is called the *ultrapower* of the standard model with respect to $\mathcal{U}$.

To prove that $\mathfrak{M}$ is a proper elementary extension of the standard model, one needs an ultrafilter that does not contain finite sets. Such ultrafilters are called *nonprincipal*. This is an important issue. To prove that nonprincipal ultrafilters exist, a version of the axiom of choice is required. See [**Doe96**, Section 4.3], in particular Exercise 122.

The set $\mathcal{U}$ in Skolem's construction is a countable ultrafilter on the set of arithmetic sets of natural numbers. As we have seen, no set theoretic assumptions are needed for its construction. The corresponding $\mathfrak{M}$ is called a *definable ultrapower* of the standard model.

The next exercise is for more advanced readers.

EXERCISE 12.7. Modify Skolem's construction to show that the standard model has $2^{\aleph_0}$ pairwise nonisomorphic simple elementary extensions.

12.4. Arithmetizing Language

Each symbol of the language of arithmetic can be assigned a single number code. In the first-order language of arithmetic there are 7 logical and auxiliary symbols, two constant symbols 0 and 1 and two binary function symbols $+$ and $\cdot$. We can assign to them consecutive codes $1, \ldots, 11$ and use Cantor's pairing to code all variables x_i by $\langle 12, i \rangle$, for i in $\mathbb{N}$. Then, each arithmetic formula φ can be represented as a finite sequence of natural numbers, and that sequence is coded by a single number via Gödel's coding. We call the code of φ the *Gödel number* of φ and denote it by $\ulcorner \varphi \urcorner$.

Now we need to revisit Definition 2.3. It describes the process of generating all first-order formulas. It is a recursive definition and it can be reformulated in terms of Gödel numbers. Then the recursive definition can be converted to a first-order arithmetic formula $\mathsf{Form}(x)$ such that for all natural numbers n, $(\mathbb{N}, +, \cdot) \models \mathsf{Form}(n)$ if and only if n is a Gödel number of a formula. It takes some skill to write this formula down, but it is fairly routine. For technical details see [**Kay91**].

If $\mathfrak{M}$ is an elementary extension of the standard model, then, by elementarity, for all standard n it still holds that $\mathfrak{M} \models \mathsf{Form}(n)$ if and only if n is a Gödel number of an arithmetic formula, but in addition there are unboundedly many nonstandard c in such that $\mathfrak{N} \models \mathsf{Form}(c)$. We can think of those c as Gödel numbers of nonstandard formulas. This idea was first considered by Abraham Robinson in [**Rob63**].

We will discuss the following arithmetized version of $\mathrm{Th}(\mathbb{N})$.

$$\mathcal{T} = \{(\ulcorner\varphi(x)\urcorner, n) : \mathfrak{N} \models \varphi(n)\}.$$

For notational convenience, the definition of $\mathcal{T}$ involves only formulas with one free variable. It is not an essential restriction. Given a sentence $\varphi(n_1, \dots, n_k)$, we can find out if it holds in the standard model, by checking whether $(\ulcorner\psi(x)\urcorner, m)$ is in $\mathcal{T}$, where $m = \langle n_1, \dots, n_k \rangle$ and $\psi(x)$ is

$$\forall y_1, \dots, \forall y_k [(y_1 = (x)_1 \wedge \cdots \wedge y_k = (x)_k) \implies \varphi(y_1, \dots, y_k)].$$

In this sense $\mathcal{T}$ codes the truth about the standard model, so we can ask whether, under this coding, the truth about the standard model is arithmetic. Tarski's theorem says that it is not. The theorem is of fundamental importance, but it is not difficult to prove. Its original proof rests on a suitable formalization of liar's paradox. It can be found in most textbooks on mathematical logic. The goal of this chapter is to give a model-theoretic proof of Tarski's theorem, due to Robinson [**Rob63**]. For the proof we will need the overspill principle.

12.5. Overspill

Every natural number can be reached from 0 by iterating the successor function finitely many times. In particular, this implies that the following induction schema of PA must hold in any structure with domain $\mathbb{N}$, i.e., for every formula $\varphi(x, y)$ of the language of the structure,

$$\forall y [\varphi(0, y) \wedge [\forall x (\varphi(x, y) \implies \varphi(x + 1, y))] \implies \forall x \varphi(x, y)].$$

EXAMPLE 12.8. Let $\mathfrak{M}$ be an elementary extension of the standard model. Consider the structure $(M, \mathbb{N})$, for which the language consists of one unary relation symbol N interpreted as $\mathbb{N}$. The induction principle fails in $\mathfrak{M}$ for the atomic formula $N(x)$. In $\mathfrak{M}$ we have $N(0)$ and for all n if $N(n)$ then $N(n+1)$, but $\mathfrak{M} \models \exists x \neg N(x)$.

If $\mathfrak{M}$ is an elementary extension of the standard model, by elementarity, the induction principle still holds in $\mathfrak{N}$ for all arithmetic formulas. From this we can derive a useful corollary that is known as the *overspill principle*.

PROPOSITION 12.9. Let $\mathfrak{M}$ be an elementary extension of the standard model, and let I be a initial segment of M that is closed under the successor function $x \mapsto x + 1$. Suppose that for a tuple $\bar{a}$ in $M^{<\omega}$ and a formula $\varphi(x, \bar{y})$, for all all i in I, $\mathfrak{M} \models \varphi(i, a)$. Then there is a b such that for all i in I, $i < b$ and $\mathfrak{M} \models \varphi(b, a)$.

PROOF. Suppose that $\mathfrak{M} \models \varphi(i, \bar{a})$ for all i in I, but for no b that is larger than all elements if I we have $\mathfrak{M} \models \varphi(b, \bar{a})$. This would mean that $\varphi(x, \bar{a})$ defines I in $\mathfrak{M}$, but that contradicts the induction principle, by the same argument as was used in Example 12.8. □

EXERCISE 12.10. Let $\mathfrak{M}$ be an ordered structure which has a least element 0 and in which every element has an immediate successor. Suppose that $\mathfrak{M}$ does not satisfy the induction schema for the language of $\mathfrak{M}$. Prove that M has a proper initial segment that is definable in $\mathfrak{M}$.

12.6. Undefinability of Truth

Tarski's theorem is more general, but we will only prove its following special case for the standard model.

THEOREM 12.11. *The set*

$$\mathcal{T} = \{ (\ulcorner \varphi(x) \urcorner, n) : \mathfrak{N} \models \varphi(n) \}$$

is not arithmetic.

PROOF. Suppose that $\mathcal{T}$ is arithmetic, i.e., there is an arithmetic formula $\Phi(x, y)$ such that for all $\varphi(x)$,

$$\mathfrak{N} \models \forall y [\Phi(\ulcorner \varphi(x) \urcorner, y) \iff \varphi(y)]. \tag{$*$}$$

Notice that the only free variable in $\Phi(\ulcorner \varphi(x) \urcorner, y)$ above is y, because $\ulcorner \varphi(x) \urcorner$ is a natural number.

Let $\mathfrak{M}$ be a simple elementary extension of the standard model, and let c be a nonstandard element of $\mathfrak{M}$ such that $\mathfrak{M} = \mathrm{Scl}(c)$.

For every b in M there is a formula $\varphi(x, y)$ such that $\mathfrak{M} \models \mu_\varphi(b, c)$, i.e., b is the least element of M for which $\mathfrak{M} \models \varphi(b, c)$ (see Definition 10.4). We will say that φ defines b from c.

For each arithmetic formula $\varphi(x, y)$, let $\nu_\varphi(x)$ be the following formula that expresses that φ does not define $(x)_1$ from $(x)_2$.

$$\forall v \forall w [(v = (x)_1 \wedge w = (x)_2) \implies \neg \mu_\varphi(v, w).]$$

The function $\ulcorner \varphi(x, y) \urcorner \mapsto \ulcorner \nu_\varphi(x) \urcorner$ is computable; hence there is an arithmetic formula $\mathsf{Form}_\nu(x)$, such that for all standard n, $\mathfrak{M} \models \mathsf{Form}_\nu(n)$ if and only if n is a Gödel number of a formula of the form $\nu_\varphi(x)$. This is by no means obvious. All details are presented in [**Kay91**].

Now we will write a formula $\Psi(z, w)$ that uses $\Phi(x, y)$ to express that there is an x that is not defined from w by any φ such that $\ulcorner \nu_\varphi \urcorner < z$.

$$\exists x \forall y [(y < z \wedge \mathsf{Form}_\nu(y)) \implies \Phi(y, \langle x, w \rangle)].$$

It may not be obvious at a first glance, but it can be checked by tracing back the definitions that for each standard n, $\mathfrak{M} \models \Psi(n, c)$; hence, by overspill, there is a nonstandard b such that $\mathfrak{M} \models \Psi(b, c)$. This means that there is an a in $\mathrm{Scl}(c)$ such that $\mathfrak{M} \models \neg \mu_\varphi(a, c)$ for all formulas $\varphi(x, y)$ whose Gödel numbers are less than b. Because b is nonstandard it means that $\mathfrak{M} \models \neg \mu_\varphi(a, c)$ holds for all formulas. This contradicts the definition of Scl(c) and finishes the proof. $\mathcal{T}$ is not arithmetic. $\qquad\qquad\qquad\qquad\qquad\qquad\qquad\qquad\qquad\qquad\qquad\square$

Tarski's theorem shows that there is no uniform arithmetically defined procedure to decide if a statement in the language of arithmetic is true in the standard model. It is even more interesting when contrasted with the fact that the truth predicate $\mathcal{T}$ becomes arithmetic when restricted to formulas with a fixed number of quantifiers.

For each n, let $\mathcal{T}_n$ be $\mathcal{T}$ restricted to formulas with n quantifiers.

THEOREM 12.12. *For each n, $\mathcal{T}_n$ is arithmetic.*

As most results mentioned in this chapter, Theorem 12.12 is more general and it is usually formulated for the hierarchy of formulas that is based on counting of alternating blocks of existential and universal quantifiers in formulas in the prenex normal form, disregarding so called bounded quantifiers. The theorem in its full generality is fundamental in model theory of arithmetic. A full discussion, with its all gory details, is included in [**Kay91**, Chapter 9].

CHAPTER 13

Saturation

In this chapter we will take a look at structures that realize as many complete types as possible. Every structure has an elementary extension that realizes all complete types that are consistent with the theory of the structure, but we want models that are richer. Together with the types in the pure language, we also want to realize types with arbitrary parameters. In this case, there must be a restriction on the cardinality of the sets of parameters in the formulas of the types. For example, for any structure $\mathfrak{M}$ with an infinite domain, the type $\{x \neq a : a \in M\}$ is consistent with $\mathrm{Th}(\mathfrak{M})$ because each of its finite fragments can be realized by an element of M, but no element of the domain realizes the type.

To construct models realizing many types, we will use elementary embeddings which are defined next.

DEFINITION 13.1. Let $\mathfrak{M}$ and $\mathfrak{N}$ be models of a theory T. A function $f : M \longrightarrow N$ is an *elementary embedding* if for all formulas φ of $\mathcal{L}_T$ and all $\bar{a}$ in $M^{<\omega}$,

$$\mathfrak{M} \models \varphi(\bar{a}) \Longleftrightarrow \mathfrak{N} \models \varphi(f(\bar{a})).$$

It follows from the definition than any elementary embedding is one-to-one. To see this, consider the formula $x = y$.

If there is an elementary embedding of $\mathfrak{M}$ into $\mathfrak{N}$, then by identifying M with its image under the embedding, $\mathfrak{N}$ becomes an elementary extension of $\mathfrak{M}$. (See the discussion following Corollary 7.8.)

DEFINITION 13.2. A type $p(\bar{x})$ in a language of a theory T is a *type of T* if $p(\bar{x})$ is realized in some model of T.

By the compactness theorem, if each finite fragment of $p(\bar{x})$ is a type of a theory T, then $p(\bar{x})$ is a type of T.

PROPOSITION 13.3. If $p(\bar{x})$ is a type of a complete theory T, then every model of T has an elementary extension in which $p(\bar{x})$ is realized.

PROOF. For a given complete theory T, let $p(\bar{x})$ be a type of T. By definition, T has a model in which $p(\bar{x})$ is realized. Let this model be $\mathfrak{N}$, and let $\bar{a}$ be a tuple that realizes $p(\bar{x})$ in $\mathfrak{N}$. Let $\mathfrak{M}$ be any other model of T.

First let us show that $p(\bar{x})$ is a type of $\mathrm{Th}(\mathfrak{M}, a)_{a \in M}$. If $\varphi_1(\bar{x})$, $\ldots$, $\varphi_n(\bar{x})$ are in $p(\bar{x})$, then

$$\mathfrak{N} \models \varphi_1(\bar{a}) \wedge \cdots \wedge \varphi_n(\bar{a}).$$

Hence,

$$\mathfrak{N} \models \exists \bar{x}[\varphi_1(\bar{x}) \wedge \cdots \wedge \varphi_n(\bar{x})].$$

Then, since T is complete, we have

$$\mathfrak{M} \models \exists \bar{x}[\varphi_1(\bar{x}) \wedge \cdots \wedge \varphi_n(\bar{x})].$$

This shows that every finite fragment of $p(\bar{x})$ is consistent with $\mathrm{Th}(\mathfrak{M}, a)_{a \in M}$, hence, by the compactness theorem, $p(\bar{x})$ is realized in a model $\mathfrak{K}$ of $\mathrm{Th}(\mathfrak{M}, a)_{a \in M}$. The function $a \mapsto a^{\mathfrak{K}}$ is an elementary embedding of $\mathfrak{M}$ into $\mathfrak{K}$ and this finishes the proof. □

EXERCISE 13.4. Let T be a complete theory. Prove that every model of T has an elementary extension which realizes all complete types of T. HINT: Use Proposition 13.3 and the elementary chain lemma (Lemma 6.11).

13.1. $\aleph_0$-saturation

Saturation is a notion of richness: the more types a structure realizes the more saturated it is. Because no model $\mathfrak{M}$ with an infinite domain can realize all types of $\mathrm{Th}(\mathfrak{M}, a)_{a \in M}$, there are different levels of saturation corresponding to various sorts of types. We will begin with a modest form of saturation which only requires that all consistent types with finitely many parameters are realized. Later, in Section 13.5 we will consider an even weaker notion.

DEFINITION 13.5. A structure $\mathfrak{M}$ is $\aleph_0$-*saturated* if for every finite $A \subseteq M$, all 1-types of $\mathrm{Th}(\mathfrak{M}, a)_{a \in A}$ are realized in $\mathfrak{M}$.

It follows from Proposition 13.3 and Theorem 6.4 that every structure with a finite domain is $\aleph_0$-saturated.

EXERCISE 13.6. Prove that all finite structures are $\aleph_0$-saturated directly from the definition.

The next proposition shows that replacing 1-types with n-types in Definition 13.5 does not result in a stronger notion of saturation.

PROPOSITION 13.7. If $\mathfrak{M}$ is $\aleph_0$-saturated then for all n and all finite $A \subseteq M$, all n-types of $\mathrm{Th}(\mathfrak{M}, a)_{a \in A}$ are realized in $\mathfrak{M}$.

PROOF. The proof is by induction on n. The case of $n = 1$ is the definition of $\aleph_0$-saturation. Assume that the proposition holds for some n. For a given $\bar{a}$ in $M^{<\omega}$, let T be $\mathrm{Th}(\mathfrak{M}, \bar{a})$. Let $p(\bar{x})$ be an $(n+1)$-type of T. We will show that $p(\bar{x})$ is realized in $\mathfrak{M}$. Let $q(x)$ be the 1-type

$$\{\exists x_1 \cdots \exists x_n \varphi(x_1, \ldots, x_n, x, \bar{a}) : \varphi(x_1, \ldots, x_n, x_{n+1}, \bar{a}) \in p(\bar{x})\}.$$

Because $p(\bar{x})$ is realized in some model of T, $q(x)$ is also realized in that model; hence, being a 1-type of T, it is realized in $\mathfrak{M}$. Pick a b that realizes $q(x)$ in $\mathfrak{M}$. Then

$$\{\varphi(x_1, \ldots, x_n, b, \bar{a}) : \varphi(x_1, \ldots, x_n, x_{n+1}, \bar{a}) \in p(\bar{x})\}$$

is an n-type of T that, by the inductive assumption, is realized in $\mathfrak{M}$ by some n-tuple $\bar{c}$. Then, $\bar{c}, b$ realizes $p(\bar{x})$. $\square$

If $\mathfrak{M}$ is $\aleph_0$-saturated, then, in particular, all n-types of $\mathrm{Th}(\mathfrak{M})$ are realized in $\mathfrak{M}$, so if $\mathfrak{M}$ is countable, for each n, there can be only countable many complete n-types of $\mathrm{Th}(\mathfrak{M})$.

EXAMPLE 13.8. Let $\mathbb{P}$ be the set of prime natural numbers. For each $X \subseteq \mathbb{P}$, let $p_X(x)$ be the type

$$\{\exists y \ (x = \underbrace{y + y + \cdots + y}_{p\text{-times}}) : p \in X\} \cup \{\forall y \neg (x = \underbrace{y + y + \cdots + y}_{q\text{-times}}) : q \in (\mathbb{P} \setminus X)\}.$$

If $p_1, \ldots, p_n$ are the primes occurring in the existential formulas of a finite fragment of $p_X(x)$, than the product $\prod_{i=1}^n p_i$ realizes all formulas in that finite fragment. Hence $p_X(x)$ is finitely realizable in $(\mathbb{N}, +)$.

Each $p_X(x)$ extends to a complete type of $\mathrm{Th}(\mathbb{N}, +)$ and there are $2^{\aleph_0}$ of them; hence, $\mathrm{Th}(\mathbb{N}, +)$ has no countable $\aleph_0$-saturated model and neither does the theory of any expansion of $(\mathbb{N}, +)$.

By Theorem 9.9, for each $n > 0$, $\aleph_0$-categorical structures realize only finitely many parameter-free complete n-types. Despite that, as the next proposition shows, they are all $\aleph_0$-saturated.

PROPOSITION 13.9. Every countable $\aleph_0$-categorical structure is $\aleph_0$-saturated.

PROOF. Let $\mathfrak{M}$ be countable and $\aleph_0$-categorical, and let $p(x, \bar{a})$, for some $\bar{a}$ in $M^{<\omega}$, be a type that is finitely realizable in $\mathfrak{M}$. Then, by Proposition 13.3, $p(x, \bar{a})$ is realized in an elementary extension $(\mathfrak{N}, \bar{a})$. By Corollary 9.17, $(\mathfrak{M}, \bar{a})$ is $\aleph_0$-categorical. Since the language of $\mathfrak{M}$ is ether finite or countable, by the Löwenheim-Skolem theorem we can assume that N is countable. By $\aleph_0$-categoricity, $(\mathfrak{N}, \bar{a})$ is isomorphic to $(\mathfrak{M}, \bar{a})$; hence $p(x, \bar{a})$ is realized in $\mathfrak{M}$. $\square$

We also have the following result, that is left as an exercise.

EXERCISE 13.10. Show that every model $\mathfrak{M}$ of an $\aleph_0$-categorical theory is $\aleph_0$-saturated (we do not assume that the model is countable). HINT: For a given 1-type $p(x, \bar{a})$, consider a countable elementary submodel of $\mathfrak{M}$ that contains $\bar{a}$.

13.1.1. More saturation. For a cardinal number κ, a structure $\mathfrak{M}$ is κ-*saturated* if for every subset A of M of cardinality less than κ, all 1-types of $\mathrm{Th}(\mathfrak{M}, a)_{a \in A}$ are realized in $\mathfrak{M}$. A structure $\mathfrak{M}$ is *saturated* if it is κ-saturated for every κ smaller than the cardinality of M.

By definition, all countable $\aleph_0$-saturated structures are saturated.

Using the results we have discussed already and a bit of set theory, it is not difficult to show that for every cardinal number κ, every structure has an elementary extension that is κ-saturated. The proof that every structure has a saturated elementary extension additionally requires a set-theoretic assumption that is not provable in ZFC, so there are complications that we will not discuss here. For some applications, saturated models are essential, but for many it is enough to work with κ-saturated models, for sufficiently large κ.

Among properties that make saturated models useful, are homogeneity and universality. We will discuss both later in this chapter, but for now let us take a look at two examples that illustrate these notions.

13.2. Types of the Theory of $(\mathbb{N}, <)$

For this section, let $T_{\mathbb{N}}$ be $\mathrm{Th}((\mathbb{N}, <))$.

The formula $\mathrm{Succ}_n(x, y)$ from Definition 4.20 is in the language of $T_{\mathbb{N}}$ and it expresses that y is the n-th successor of x. For each n in $\mathbb{N}$, $p(x) = \mathrm{tp}^{(\mathbb{N}, <)}(n)$ is isolated by the formula $\forall z[\forall w \neg (w < z) \implies \mathrm{Succ}_n(z, x)]$.

Because the all complete 1-types realized in $(\mathbb{N}, <)$ are isolated, all complete n-types realized in $(\mathbb{N}, <)$ are isolated as well. For example, if $\varphi(x)$ isolates $\mathrm{tp}^{(\mathbb{N}, <)}(m)$ and $\psi(x)$ isolates $\mathrm{tp}^{(\mathbb{N}, <)}(n)$, then $\varphi(x) \wedge \psi(y)$ isolates $\mathrm{tp}^{(\mathbb{N}, <)}(m, n)$. It does not work this way in general, usually there is much more that we need to know the type of a pair of elements given the types of each element in the pair. We will see examples soon. Here, we take advantage of the fact that each n is uniquely determined by its type in $(\mathbb{N}, <)$.

We know all there is to know about the types realized in $(N, <)$, but what about the complete n-types that realized in other models of $T_{\mathbb{N}}$? For this discussion, we need a few definitions.

Addition of ordered sets has already been introduced in Chapter 9. We will now do it more formally.

Let $(A, <)$, $(B, <)$ be ordered sets. Then the sum of $(A, <)$ and $(B, <)$ is the ordered set $(A \oplus B, <)$ whose domain is $(\{0\} \times A) \cup (\{1\} \times B)$, with the ordering defined by

$$(i, a) < (j, b) \text{ iff either } i < j, \text{ or } i = j \text{ and } a < b.$$

We can also multiply ordered sets. Think of the product $(A \otimes B, <)$ as the set obtained by replacing each a in A by a copy of $(B, <)$. Formally, $(A \otimes B, <)$ is the ordered set with the domain $A \times B$ with the ordering defined by $(a_1, b_1) < (a_2, b_2)$ if and only if either $a_1 < a_2$, or $a_1 = a_2$ and $b_1 < b_2$.

EXERCISE 13.11. Show that $(A \oplus A, <)$ is isomorphic to $(I \otimes A, <)$, where I is $\{0, 1\}$ with the usual ordering.

EXERCISE 13.12. Let $(M, <)$ be an elementary extension of $(\mathbb{N}, <)$. Show that there is an ordered set $(A, <)$ such that $(M, <)$ is isomorphic to $(\mathbb{N} \oplus (A \otimes \mathbb{Z}), <)$.

The proof of the following theorem involves Ehrenfeucht-Fraïsse games—a technique that will not be covered in these lectures. See [**Doe96**, Chapter 3] or [**Mar02**, Section 2.4].

THEOREM 13.13. *For any ordered sets $(I, <)$, $(J, <)$, if $I \subseteq J$, then*

$$(\mathbb{N} \oplus (I \otimes \mathbb{Z}), <) \prec (\mathbb{N} \oplus (J \otimes \mathbb{Z}), <).$$

In particular, $(\mathbb{N} \oplus \mathbb{Z}), <)$ is an elementary extension of $(\mathbb{N}, <)$.

Equipped with Theorem 13.13, we will now examine some types of $T_{\mathbb{N}}$ to determine which of the elementary extensions of $(\mathbb{N}, <)$ given by Theorem 13.13 are $\aleph_0$-saturated.

By Proposition 13.7, we can prove that a model is not $\aleph_0$-saturated, by finding an n-type with finitely many parameters that is not realized in the model. The discussion will also lead us to a classification of all complete types of $T_{\mathbb{N}}$.

The type

$$p_1(x_1) = \{\forall y[\mathrm{Succ}_n(0, y) \Longrightarrow y < x_1] : n \in \mathbb{N}\}$$

is not realized in $(\mathbb{N}, <)$, but it is realized by any nonstandard element in $(\mathbb{N} \oplus \mathbb{Z}, <)$.

The type

$$p_2(x_1, x_2) = \{\forall y[\mathrm{Succ}_n(0, y) \Longrightarrow y < x_1] \wedge \forall y[\mathrm{Succ}_n(x_1, y) \Longrightarrow y < x_2] : n \in \mathbb{N}\}$$

is not realized in $(\mathbb{N} \oplus \mathbb{Z}, <)$, but it is realized in $(\mathbb{N} \oplus \mathbb{Z} \oplus \mathbb{Z}, <))$.

For each $n > 1$, there is a type $p_n(x_1, \ldots, x_n)$ expressing that x_1 is non-standard and that the sequence $x_1, \ldots, x_n$ is increasing, with terms that are infinitely far apart. To realize this type, we need $\mathbb{N}$ followed by at least n copies of $\mathbb{Z}$.

For each n, $p_n(x_1, \ldots, x_n)$ is realized in $(\mathbb{N} \oplus (\mathbb{N} \otimes \mathbb{Z}), <)$. In fact, we will show now that all n-types of $T_{\mathbb{N}}$ are realized in $(\mathbb{N} \oplus (\mathbb{N} \otimes \mathbb{Z}))$.

Let $p(x_1, \ldots, x_n)$ be a type of $T_{\mathbb{N}}$ and let $(\mathbb{N} \oplus (I \otimes Z))$ be a model of $T_{\mathbb{N}}$ in which $p(x_1, \ldots, x_n)$ is realizes by a tuple $a_1, \ldots, a_n$. Let J be a finite subset of I such that all $a_1, \ldots, a_n$ are in $(\mathbb{N} \oplus (J \otimes \mathbb{Z})$. It follows from Theorem 13.13 that $p(x_1, \ldots, x_n)$ is realized in $(\mathbb{N} \oplus (J \otimes \mathbb{Z}), <))$ and, because $(\mathbb{N} \oplus (J \otimes \mathbb{Z}), <))$ is isomorphic to an elementary submodel of $(\mathbb{N} \oplus (\mathbb{N} \otimes \mathbb{Z}), <)$, this proves our claim.

Recall that a $\mathbb{Z}$-block of $(\mathbb{N} \oplus (I \otimes \mathbb{Z}), <)$ is the set $\{a + n : n \in \mathbb{Z}\}$ for some a in $I \otimes \mathbb{Z}$.

We have shown that $(\mathbb{N} \oplus (\mathbb{N} \otimes \mathbb{Z}), <)$ realizes all types of $T_{\mathbb{N}}$, but still $(\mathbb{N} \oplus (\mathbb{N} \otimes \mathbb{Z}), <)$ is not $\aleph_0$-saturated. To see this, let a be an element of the least $\mathbb{Z}$-block in $\mathbb{N} \otimes \mathbb{Z}$. For each n, there is a formula $\varphi_n(x, y)$ of the language of $T_{\mathbb{N}}$ expressing that $n < x < y - n$. The type $\{\varphi_n(x, a) : n \in \mathbb{N}\}$ is not realized in $(\mathbb{N} \oplus (\mathbb{N} \otimes \mathbb{Z}), <)$. We can remedy this by moving to $(\mathbb{N} \oplus (\mathbb{Z} \otimes \mathbb{Z}), <)$, but this one is also not $\aleph_0$-saturated. To prove it, let a and b be in two adjacent $\mathbb{Z}$-blocks, and suppose that $a < b$. For each n, let $\psi_n(x, y, z)$ be a formula of the language of $T_\mathbb{N}$ expressing that $y + n < x < z - n$. Then the type $\{\psi_n(x, a, b) : n \in \mathbb{N}\}$ is not realized in $(\mathbb{N} \oplus (\mathbb{Z} \otimes \mathbb{Z}), <)$.

The discussion above shows that if $(\mathbb{N} \oplus (I \oplus \mathbb{Z}), <)$ is $\aleph_0$-saturated, then $(I, <)$ must be densely ordered without end points. It is also a sufficient condition.

EXERCISE 13.14. Use Exercise 13.12, Theorem 13.13, and Exercise 5.12 to show that $(\mathbb{N} \oplus (\mathbb{Q} \otimes \mathbb{Z}), <)$ is $\aleph_0$-saturated.

To classify all complete types realized in models of $T_\mathbb{N}$, we will use the following *distance function*

$$d(a, b) = \begin{cases} 0, & \text{if } a = b, \\ n, & \text{if } \mathrm{Succ}_n(a, b) \\ \infty, & \text{otherwise.} \end{cases}$$

Let $\mathfrak{M}$ and $\mathfrak{N}$ be countable models of $T_\mathbb{N}$, and let $\bar{a} = a_1, \ldots, a_n$ in $M^{<\omega}$ and $\bar{b} = b_1, \ldots, b_n$ in $N^{<\omega}$ be increasing. It can be shown directly, but it also follows from Theorem 13.24 below, that $\mathfrak{M}$ and $\mathfrak{N}$ can be elementarily embedded into $(\mathbb{N} \oplus (\mathbb{Q} \otimes \mathbb{Z}), <)$. Let f and g be such embeddings. It is easy to see that if $d(0, a_1) = d(0, b_1)$ and for all i, $d(a_i, a_{i+1}) = d(b_i, b_{i+1})$, then there is an α in $\mathrm{Aut}(\mathbb{N} \oplus (\mathbb{Q} \otimes \mathbb{Z}), <)$ such that $\alpha(f(\bar{a})) = \alpha(g(\bar{b}))$; hence, $\mathrm{tp}(\bar{a}) = \mathrm{tp}(\bar{b})$.

We have obtained a classification of all complete n-types of $T_\mathbb{N}$. The type of each increasing n-tuple is completely determined by the distances between consecutive elements and the distance of the least element from 0. If all those distances are finite, then the type of the tuple is isolated. If at least one of the distances is infinite, the type is not isolated. To see this, let $\bar{a}$ be a tuple in $(\mathbb{N} \oplus (I \otimes \mathbb{Z}), <)$ such that at least one distance between its elements is infinite. Let $\varphi(\bar{x})$ be a formula in $\mathrm{tp}(\bar{a})$. Then

$$(\mathbb{N} \oplus (I \otimes \mathbb{Z})), <) \models \exists \bar{x} \, \varphi(\bar{x}),$$

and then, by Theorem 13.13,

$$(\mathbb{N}, <) \models \exists \bar{x} \, \varphi(\bar{x}).$$

Since $\mathrm{tp}(\bar{a})$ is not realized in $(\mathbb{N}, <)$, this shows that $\varphi(\bar{x})$ does not isolate $\mathrm{tp}(\bar{a})$.

13.3. Definitional Reducts

In this section, we will take a look at an application of a special kind of reduct. It illustrates a simple, but very useful technique.

The complete theory of $(\mathbb{N}, <)$ is particularly simple and its models are easy to classify. In this section we will examine an even simpler structure. Let S be the successor relation on the set of natural numbers, i.e., for all m and n in $\mathbb{N}$, (m, n) is in S if an only of $n = m + 1$. We could analyze the models of $\mathrm{Th}((\mathbb{N}, S))$ directly, as we did it for $\mathrm{Th}((\mathbb{N}, <))$ in the previous section, but we will take a different route, taking advantage of the fact that S is definable in

$(\mathbb{N}, <)$. This gives us an opportunity to introduce the concept of definitional reduct.

Let $\mathfrak{M}$ be a structure and let $\mathfrak{X}$ be a set of parametrically definable relations of $\mathfrak{M}$. The structure $(M, \mathfrak{X})$ is called a *definitional reduct* of $\mathfrak{M}$. Notice the use of font, we are not talking about $(\mathfrak{M}, \mathfrak{X})$. The only relations of $(M, \mathfrak{X})$ are those in $\mathfrak{X}$; all other relations, functions, and constants of $\mathfrak{M}$ have been eliminated. This justifies calling $(M, \mathfrak{X})$ a reduct.

For every formula $\varphi(\bar{x})$ of the language of $(M, \mathfrak{X})$ there is a formula $\varphi^*(\bar{x})$ of the language of $\mathfrak{M}$ in which every occurrence of a relation symbol is replaced by a formula that defines the relation in $\mathfrak{M}$. Then, for all $\bar{a}$ in M

$$(M, \mathfrak{X}) \models \varphi(\bar{a}) \text{ iff } \mathfrak{M} \models \varphi^*(\bar{a}).$$

It follows that a definitional reduct of an $\aleph_0$-saturated structure is $\aleph_0$-saturated.

Definitional reduct can be similarly defined for $\mathfrak{X}$ that also includes parametrically definable functions and constants, but then the translation process is a bit more elaborate. For example, suppose that c is defined by in $\mathfrak{M}$ by $\psi(x)$ and $f : M \longrightarrow M$ is defined by $\theta(x, y)$. Let $\varphi(x)$ be $f(x) = c$. Then $\varphi^*(x)$ is

$$\forall y, z[(\psi(z) \wedge \varphi(x, y)) \implies y = z].$$

EXERCISE 13.15. Prove that a definitional reduct of an $\aleph_0$-categorical structure is $\aleph_0$-categorical. HINT: Use Theorem 9.9.

The successor relation on the set of natural numbers is definable in $(\mathbb{N}, <)$: $n = m + 1$ if and only if $(\mathbb{N}, <) \models \text{Succ}(m, n)$, where $\text{Succ}(x, y)$ is the formula $x < y \wedge \forall z[x < z \implies (y = z \vee y < z)]$.

As before, let $T_{\mathbb{N}}$ be $\text{Th}((\mathbb{N}, <))$. For a model $\mathfrak{M}$ of $T_{\mathbb{N}}$, let

$$S^{\mathfrak{M}} = \{(m, n) : (m, n) \in M^2 \text{ and } \mathfrak{M} \models \text{Succ}(m, n)\}.$$

For each model $\mathfrak{M}$ of $T_{\mathbb{N}}$, $(M, S^{\mathfrak{M}})$ is a definitional reduct of $\mathfrak{M}$.

Let T_S be $\text{Th}(\mathbb{N}, S)$. We will use S both as the name of the successor relation on $\mathbb{N}$ and as the relation symbol of T_S. So if $\mathfrak{M}$ is a model of T_S, then $S^{\mathfrak{M}}$ is the interpretation of S in $\mathfrak{M}$. The following statements hold in $(\mathbb{N}, S)$; hence also in every model $\mathfrak{M}$ of T_S.

$$\exists x \forall y \neg S(y, x). \tag{1}$$

Let $0^{\mathfrak{M}}$ be the unique element of M given by (1).

$$\forall x[\neg(x = 0^{\mathfrak{M}}) \implies (\exists y, z(S(y, x) \wedge S(x, z)))], \tag{2}$$

and for each n,

$$\neg \exists x_0 \ldots, x_n[\bigwedge \{S(x_i, x_{i+1}) : i < n\} \wedge S(x_n, x_0)]. \tag{3}$$

$0^{\mathfrak{M}}$ is definable in $\mathfrak{M}$ and so is its successor, call it $1^{\mathfrak{M}}$, and the successor of $1^{\mathfrak{M}}$, call it $2^{\mathfrak{M}}$, and so on. So $\mathfrak{M}$ contains an isomorphic copy of $(\mathbb{N}, S)$. Let us identify this copy with $(\mathbb{N}, S)$.

Let us call a model isomorphic to $(\mathbb{N}, S)$ an $\mathbb{N}$-chain and a model isomorphic to $(\mathbb{Z}, S)$ a $\mathbb{Z}$-chain. It follows from (2) and (3) that if a model $\mathfrak{M}$ of T_S is not isomorphic to $(\mathbb{N}, S)$, then each element of $M \setminus \mathbb{N}$ belongs to a $\mathbb{Z}$-chain. Hence M is the union of an $\mathbb{N}$-chain and the union of a set of disjoint $\mathbb{Z}$-chains.

Models of T_S are not ordered, but any linear ordering of the $\mathbb{Z}$-chains of a model of T_S can be used to define an expansion of a model of T_S to a model of $T_{\mathbb{N}}$. This observation can be used to do the following exercise that summarizes this section.

EXERCISE 13.16. Let $\mathfrak{M}$ be a countable model of T_S. Prove that $\mathfrak{M}$ is $\aleph_0$-saturated if and only if $\mathfrak{M}$ has infinitely many $\mathbb{Z}$-chains. HINT: If $\mathfrak{M}$ has infinitely many $\mathbb{Z}$-chains, they can be ordered densely.

13.4. Universality and Homogeneity

Let $\mathfrak{M} = (\mathbb{N} \oplus (\mathbb{N} \otimes \mathbb{Z}), <)$. In Section 13.2, we have seen that all n-types of $\mathrm{Th}((\mathbb{N}, <))$ are realized in $\mathfrak{M}$, but $\mathfrak{M}$ is not $\aleph_0$-saturated. One obstacle is the lack of homogeneity (see Definition 9.11). If A, B, and C are consecutive $\mathbb{Z}$-blocks of $\mathfrak{M}$ and $a \in A$, $b \in B$, and $c \in C$, then $\mathrm{tp}(a, b) = \mathrm{tp}(a, c)$, but there is no d such that $\mathrm{tp}(a, c, b) = \mathrm{tp}(a, b, d)$. In other words, in this example $\mathrm{Th}(\mathfrak{M}, a, b) = \mathrm{Th}(\mathfrak{M}, a, c)$, but $(\mathfrak{M}, a, b)$ is not isomorphic to $(\mathfrak{M}, a, c)$. The pair (a, b) has all the same first-order properties that (a, c) has, but we can tell a difference between them that is not first-order. For example, a and c are separated by a $\mathbb{Z}$-block, but a and b are not.

EXERCISE 13.17. Let $\mathfrak{M} = (\mathbb{N} \oplus (\mathbb{N} \otimes \mathbb{Z}), <)$. Find a and b such that $\mathrm{tp}^{\mathfrak{M}}(a) = \mathrm{tp}^{\mathfrak{M}}(b)$ and (M, a) is not isomorphic to (M, b).

It turns out that for models realizing all types without parameters, the lack of homogeneity is the only obstacle.

THEOREM 13.18. *A model of a complete theory T is $\aleph_0$-saturated if and only if it is homogeneous and it realizes all n-types of T, for each $n > 0$.*

PROOF. Let $\mathfrak{M} \models T$ be $\aleph_0$-saturated, and let $p(\bar{x})$ be an n-type realized in a model $\mathfrak{N}$ of T. Since T is complete, $\mathrm{Th}(\mathfrak{M}) = \mathrm{Th}(\mathfrak{N})$; hence $p(\bar{x})$ is a type of $\mathrm{Th}(\mathfrak{M})$. By Proposition 13.7, $p(\bar{x})$ is realized in $\mathfrak{M}$.

To show that $\mathfrak{M}$ is homogeneous, suppose that for some $\bar{a}$, $\bar{b}$ in $M^{<\omega}$, $\mathrm{tp}(\bar{a}) = \mathrm{tp}(\bar{b})$, and let a in M be given. Consider the type

$$p(x, \bar{b}) = \{\varphi(\bar{b}, x) : \mathfrak{M} \models \varphi(\bar{a}, a)\}.$$

Then, for any $\varphi_1(\bar{b}, x), \dots, \varphi_n(\bar{b}, x)$ in $p(x, \bar{b})$, a witnesses that

$$\mathfrak{M} \models \exists x[\varphi_1(\bar{a}, x) \wedge \cdots \wedge \varphi_n(\bar{a}, x)].$$

Since $\bar{a}$ and $\bar{b}$ have the same type,

$$\mathfrak{M} \models \exists x[\varphi_1(\bar{b}, x) \wedge \cdots \wedge \varphi_n(\bar{b}, x)].$$

This proves that $p(x, \bar{b})$ is finitely realizable, so, by $\aleph_0$-saturation, it is realized by some b in $\mathfrak{M}$. For any such b, $\mathrm{tp}(\bar{a}, a) = \mathrm{tp}(\bar{b}, b)$.

Now suppose that $\mathfrak{M}$ is homogeneous and realizes all n-types of T. Let $p(x, \bar{a})$ be a type of $\mathrm{Th}(\mathfrak{M}, \bar{a})$ for some n-tuple $\bar{a}$. Consider the $(n + 1)$-type

$$q(x, \bar{y}) = \{\varphi(x, \bar{y}) : \varphi(x, \bar{a}) \in p(x, \bar{a})\} \cup \{\psi(\bar{y}) : \mathfrak{M} \models \psi(\bar{a})\}.$$

Any finite subset of $q(x, \bar{y})$ is realized in $\mathfrak{M}$; hence, by the assumption, the whole type is realized in $\mathfrak{M}$ by some $(n + 1)$-tuple $c, \bar{b}$. Because $\mathfrak{M}$ is homogeneous and $\mathrm{tp}(\bar{a}) = \mathrm{tp}(\bar{b})$, there is a b such that $\mathrm{tp}(b, \bar{a}) = \mathrm{tp}(c, \bar{b})$. Any such b realizes $p(x, \bar{a})$. □

COROLLARY 13.19. *If $\mathfrak{M}$ and $\mathfrak{N}$ are countable, $\aleph_0$-saturated and $\mathrm{Th}(\mathfrak{M}) = \mathrm{Th}(\mathfrak{N})$, then $\mathfrak{M}$ is isomorphic to $\mathfrak{N}$.*

PROOF. This is another example of a proof by a "back-and-forth" construction.

Let $\mathfrak{M}$ ad $\mathfrak{N}$ be countable, $\aleph_0$-saturated, and suppose that $\mathrm{Th}(\mathfrak{M}) = \mathrm{Th}(\mathfrak{N})$. By Theorem 13.18, both models are homogeneous and they realize the same types. Let $M = \{a_n : n \in \mathbb{N}\}$ and $N = \{b_n : n \in \mathbb{N}\}$. By induction, we will define $\{c_n : n \in \mathbb{N}\} = M$ and $\{d_n : n \in \mathbb{N}\} = N$ so that the function $c_n \mapsto d_n$ is an isomorphism $f : \mathfrak{M} \longrightarrow \mathfrak{N}$.

To begin, let $c_0 = a_0$. Because $\mathfrak{M}$ and $\mathfrak{N}$ realize the same types, $\mathfrak{N}$ has an element realizing the type of c_0. Let d_0 be such an element.

Suppose that we have $\bar{c} = c_0, \ldots, c_n$, and $\bar{d} = d_0, \ldots, d_n$, such that $\mathrm{tp}(\bar{c}) = \mathrm{tp}(\bar{d})$ and n is even. We let c_{n+1} to be the first a_i that has not been used in the construction yet. Since $\mathrm{tp}(\bar{c}) = \mathrm{tp}(\bar{d})$, the following 1-type is finitely realizable in $\mathfrak{N}$:

$$p(x, \bar{d}) = \{\varphi(\bar{d}, x) : \mathfrak{M} \models \varphi(\bar{c}, c_{n+1})\}.$$

By $\aleph_0$-saturation, $p(x, \bar{d})$ is realized in $\mathfrak{N}$ and we let d_{n+1} to be any element of N realizing $p(x, \bar{d})$.

If n is odd we proceed the same way, but we begin by first selecting a d_{n+1} as the first b_i that has not been used in the construction yet and then finding the required c_{n+1} in M exactly as above. □

Let us go back to Theorem 13.18. In the proof, for given $\bar{a}$, $\bar{b}$ and a such that $\mathrm{tp}^{\mathfrak{M}}(\bar{a}) = \mathrm{tp}^{\mathfrak{M}}(\bar{b})$, we showed that the type

$$p(x, \bar{b}) = \{\varphi(\bar{b}, x) : \mathfrak{M} \models \varphi(\bar{a}, a)\}$$

is finitely realized in $\mathfrak{M}$; hence, by the assumption, it is realized in $\mathfrak{M}$, proving that $\aleph_0$-saturated structures are homogeneous. If $\mathfrak{M}$ is not $\aleph_0$-saturated, the argument shows that $\mathfrak{M}$ has an elementary extension $\mathfrak{N}$ such that $\mathrm{tp}^{\mathfrak{N}}(\bar{a}, a) = \mathrm{tp}^{\mathfrak{N}}(\bar{b}, b)$, for some b in N. This remark is the key to the following theorem.

THEOREM 13.20. *Every structure with a countable domain has a countable homogeneous elementary extension.*

PROOF. Let $\mathfrak{M}$ be a structure with a countable domain M. Let $\{(\bar{a}_n, \bar{b}_n, c_n) : n \in \mathbb{N}\}$ be an enumeration of all triples $(\bar{a}, \bar{b}, c)$ such that $\mathrm{tp}^M(\bar{a}) = \mathrm{tp}^M(\bar{b})$, where $\bar{a}$ and $\bar{b}$ are in $M^{<\omega}$ and c is an arbitrary element of M. Using the remark before the theorem, we now build a chain of elementary extensions

$$\mathfrak{M} \prec \mathfrak{M}_1 \prec \mathfrak{M}_2 \prec \cdots,$$

such that for each n there is a d_n in M_{n+1} such that $\mathrm{tp}^{\mathfrak{M}_{n+1}}(\bar{a}, c_n) = \mathrm{tp}^{\mathfrak{M}_{n+1}}(\bar{b}, d_n)$.

Let $\mathfrak{N}_1$ be $\bigcup_{n\in\mathbb{N}} \mathfrak{M}_n$. By the elementary chain lemma (Lemma 6.11), $\mathfrak{N}_1$ is an elementary extension of $\mathfrak{M}$.

$\mathfrak{N}_1$ may not be homogeneous yet, but we can repeat the entire process described above starting with $\mathfrak{N}_1$ instead of $\mathfrak{M}$, to obtain $\mathfrak{N}_2$, and then we keep going in this fashion to get the chain of elementary extensions

$$\mathfrak{N}_1 \prec \cdots \prec \mathfrak{N}_n \prec \cdots.$$

It is easy to check that $\mathfrak{N} = \bigcup_{n\in\mathbb{N}} \mathfrak{N}_n$ is a countable, homogeneous model of T. $\qquad\square$

As a corollary, we get the next theorem that characterizes theories that have countable $\aleph_0$-saturated models.

THEOREM 13.21. *Let T be a theory in a countable language. Then T has a countable $\aleph_0$-saturated model if T has at most $\aleph_0$ complete 1-types.*

PROOF. If $\mathfrak{M}$ is a countable $\aleph_0$-saturated model of T, then all types of T are realized in $\mathfrak{M}$, so T has at most $\aleph_0$ complete types, so we only need to prove the converse.

Let $\mathfrak{M}$ be a countable model ot T, and let $\{p_n(\bar{x}) : n \in \mathbb{N}\}$ be an enumeration of all complete types of T. First, starting with any countable model $\mathfrak{M}$ of T, using Proposition 13.3, we build a chain of elementary extensions $\mathfrak{M} \prec \mathfrak{M}_1 \prec \mathfrak{M}_2 \cdots$, such that $p_n(\bar{x})$ is realized in $\mathfrak{M}_{n+1}$. Then $\mathfrak{N} = \bigcup_{n\in\mathbb{N}} \mathfrak{M}_n$ is countable and realizes all complete types of T. By Theorem 13.20, $\mathfrak{N}$ has a homogeneous countable elementary extension $\mathfrak{K}$. By Theorem 13.18, $\mathfrak{K}$ is $\aleph_0$-saturated. $\qquad\square$

A simple example of a structure that is homogeneous but not $\aleph_0$-saturated is $(\mathbb{N}, <)$. $(\mathbb{N}, <)$ is homogeneous, for the simple reason that there are no distinct $\bar{a}$, $\bar{b}$, such that $\mathrm{tp}(\bar{a}) = \mathrm{tp}(\bar{b})$. A more interesting example is in the next exercise.

EXERCISE 13.22. Show that $(\mathbb{N} \oplus \mathbb{Z}, <)$ is homogeneous.

The structure in the exercise above is too small to be $\aleph_0$-saturated. Our last result shows that to be $\aleph_0$-saturated, a structure has to be really large in the sense of the following definition.

DEFINITION 13.23. A structure $\mathfrak{M}$ is $\aleph_0$-*universal* if for any countable $\mathfrak{N}$ that is elementarily equivalent to $\mathfrak{M}$ there is an elementary embedding of $\mathfrak{N}$ into $\mathfrak{M}$.

By performing the construction in the proof of Corollary 13.19 only in the "forth" direction, we get another corollary.

COROLLARY 13.24. $\aleph_0$-saturated structures are $\aleph_0$-universal.

EXERCISE 13.25. Prove Corollary 13.24.

It follows directly from the definition that if $\mathfrak{M}$ is $\aleph_0$-saturated, then so is $(\mathfrak{M}, \bar{a})$ for any $\bar{a}$ in $M^{<\omega}$. This simple observation gives us a short proof of the following result, that also follows from Theorems 9.13 and 13.18.

COROLLARY 13.26. Countable $\aleph_0$-saturated structures are strongly homogeneous.

PROOF. Let $\mathfrak{M}$ be countable, $\aleph_0$-saturated, and suppose that for $\bar{a}$, $\bar{b}$ in $M^{<\omega}$, $\operatorname{tp}(\bar{a}) = \operatorname{tp}(\bar{b})$. Then $(\mathfrak{M}, \bar{a})$ and $(\mathfrak{M}, \bar{b})$ are $\aleph_0$-saturated models of $\operatorname{Th}(\mathfrak{M}, \bar{a})$; hence they are isomorphic. Any isomorphism between them is an automorphism that maps $\bar{a}$ to $\bar{b}$. $\square$

$\aleph_0$-universality does not imply $\aleph_0$-saturation. For example, every countable model of $\operatorname{Th}((\mathbb{N}, <))$ elementarily embeds into $(\mathbb{N} \oplus (\mathbb{Q} \otimes \mathbb{Z}) \oplus \mathbb{Z}), <)$, but this model is not $\aleph_0$-saturated.

EXERCISE 13.27. Show that every countable $\aleph_0$-universal model of $\operatorname{Th}((\mathbb{N}, S))$, where S is the successor relation, is $\aleph_0$-saturated.

13.5. Recursive Saturation

Many theories do not have countable $\aleph_0$-saturated models. However, there is a larger class of structures that satisfy a weaker notion of saturation, in which every theory in finite language has a model, and for which many results analogous to those discussed in the previous section hold. To define the notion we need to refer to arithmetization of syntax that was discussed in Chapter 12.

In this section, we are assuming that the languages of theories are finite. Let $\mathcal{L}$ be such a language. We will also assume that the domains of all structures are infinite.

We will say that a type of $\mathcal{L}$ is *computable*, if the set of Gödel numbers of the formulas in the type is computable (see Section 12.4). If $p(\bar{x})$ is a type of $\operatorname{Th}(\mathfrak{M}, \bar{a})$, for some $\bar{a}$ in $M^{<\omega}$, then we call it a computable type, if it is a computable type of $\mathcal{L}$ extended by adding constant symbols for the elements in $\bar{a}$.

To simplify some statements and arguments, we have assumed that all languages in this sections are finite. With a bit more care, all results hold for theories in computable languages.

Computability here refers to the notion of computability that was originally defined in recursion theory. Since the 1990's, the terminology has changed and for over 30 years now the term *recursive* has been replaced by *computable* and *recursion theory* became *computability theory*. For our discussion, we will not need a formal notion of computability. We will just apply a simple principle:

if one can verify whether a formula is in a given type just by recognizing its syntactic form, then that type is computable. Examples will be given soon.

Recursive saturation was introduced by Jon Barwise and John Schlipf in the 1970's and has found numerous applications. For examples see [**CK90**, Chapter 2].

DEFINITION 13.28. A structure $\mathfrak{M}$ is *recursively saturated* if for every finite $I \subseteq M$, all *computable* 1-types of $\mathrm{Th}(\mathfrak{M}, a)_{a \in I}$ are realized in $\mathfrak{M}$.

It follows directly from the definition that if $\mathfrak{M}$ is recursively saturated, then so is $(\mathfrak{M}, \bar{a})$ for any $\bar{a}$ in $M^{<\omega}$.

In some recent publications recursive saturation has been renamed computable saturation, but we will not follow that trend. The term "computable saturation" could be used for saturated structures for which there is a computable procedure that to a computable type assigns an element that realizes it. Such procedures exist only in very special cases. We will use a hybrid terminology: recursive sets and functions will be called computable, but recursive saturation will stay recursive saturation.

The key difference between $\aleph_0$-saturation and recursive saturation is that while a theory may have uncountably many types, for any theory T in a finite or countable language there are only countably many computable types of T, because there are only countably many computable sets.

Clearly, each $\aleph_0$-saturated structure is recursively saturated, but there are many more. It follows from the next theorem.

THEOREM 13.29. *Every countable model of a theory in a finite language has a countable recursively saturated elementary extension.*

PROOF. Let $\mathfrak{M}$ be countable. By Proposition 13.3, for each $\bar{a}$ in $M^{<\omega}$ and each 1-type $p(x)$ of $\mathrm{Th}(\mathfrak{M}, \bar{a})$, $\mathfrak{M}$ has an elementary extension $\mathfrak{M}_1$ in which $p(x)$ is realized. If $q(x)$ is a 1-type of $\mathrm{Th}(\mathfrak{M}, \bar{b})$, for a $\bar{b}$ in $M^{<\omega}$, then it is also a type of $\mathrm{Th}(\mathfrak{M}_1, \bar{b})$ and it is realized in some elementary extension $\mathfrak{M}_2$ of $\mathfrak{M}_1$, so now both $p(x)$ and $q(x)$ are realized in $\mathfrak{M}_2$.

Because M is countable and there are only countable many computable sets, there is a list $\{p_n(x) : n \in \mathbb{N}\}$ of all computable 1-types of $Th(\mathfrak{M}, \bar{a})$, for all $\bar{a}$ in $M^{<\omega}$, and we can proceed as above to build a countable chain of elementary extensions

$$\mathfrak{M} \prec \mathfrak{M}_1 \prec \cdots \prec \mathfrak{M}_n \prec \cdots,$$

such that $p_1(x), \ldots, p_n(x)$ are realized in $\mathfrak{M}_n$. Let $\mathfrak{N}_1 = \bigcup_{n \in \mathbb{N}} \mathfrak{M}_n$. By the elementary chain lemma (Lemma 6.11), $\mathfrak{N}_1$ is an elementary extension of $\mathfrak{M}$ and it realizes all computable 1-types with parameters from M.

$\mathfrak{N}_1$ may not be recursively saturated yet, but iterating the entire process described above, we can build another countable chain of elementary extensions

$$\mathfrak{N}_1 \prec \cdots \prec \mathfrak{N}_n \prec \cdots,$$

such that $\mathfrak{N}_{n+1}$ realizes all computable 1-types with parameters from N_n. It is easy to check that $\mathfrak{N} = \bigcup_{n \in \mathbb{N}} \mathfrak{N}_n$ is a countable, recursively saturated elementary extension of $\mathfrak{M}$. □

With a bit more of set theory, Theorem 13.29 can be generalized to: Every model has a recursively saturated elementary extension of the same cardinality.

The proof of the next theorem is a replica of the proof of the corresponding result for $\aleph_0$-saturated models, with small modifications. A complete proof is included to show how computable types come up.

THEOREM 13.30. *Recursively saturated models are homogeneous.*

PROOF. Let $\mathfrak{M}$ be recursively saturated. Suppose that for some $\bar{a}$, $\bar{b}$ in $M^{<\omega}$, $\mathrm{tp}(\bar{a}) = \mathrm{tp}(\bar{b})$, and let a in M be given. Consider the 1-type

$$p(x, \bar{a}, \bar{b}, a) = \{\varphi(\bar{b}, x) \Longleftrightarrow \varphi(\bar{a}, a) : \varphi \in \mathcal{L}\}.$$

Digression: Notice the difference. In the proof of Theorem 13.18 we have used the type

$$p(x, \bar{b}) = \{\varphi(\bar{b}, x) : \mathfrak{M} \models \varphi(\bar{a}, a)\}.$$

To decide whether $\varphi(\bar{b}, x)$ is in $p(x, \bar{b})$ or not, one has to know whether $\mathfrak{M} \models \varphi(\bar{a}, a)$ and there may not be a computable procedure to do that. Hence the type may not be computable. The type $p(x, \bar{a}, \bar{b}, a)$ as defined in this proof is computable. It consists of formulas $\varphi(\bar{b}, x) \Longleftrightarrow \varphi(\bar{a}, a)$, for all formulas $\varphi(\bar{y}, x)$ of the language. To decide if a formula is in it or not, it is enough to look at its shape (a spell checker could do it). In 13.18, we could have defined the type as we did in this proof, but the definition used there was more convenient for the argument. The end of digression, we are returning to the proof.

Exactly as in the proof of Theorem 9.11, one can show that $p(x, \bar{a}, \bar{b}, a)$ is finitely realized in $\mathfrak{M}$, so, by recursive saturation, it is realized by some b in $\mathfrak{M}$. For any such b, $\mathrm{tp}(\bar{a}, a) = \mathrm{tp}(\bar{b}, b)$. □

Now we can apply a back-and-forth construction to prove the next corollary.

COROLLARY 13.31. *Countable recursively saturated models are strongly homogeneous.*

EXERCISE 13.32. Prove Corollary 13.31.

Countable $\aleph_0$-saturated models are unique up to isomorphism. It is not so for recursively saturated models. As we have seen in Example 13.8, $\mathrm{Th}(\mathbb{N}, +)$ has $2^{\aleph_0}$ complete types. Any such theory T has $2^{\aleph_0}$ countable nonisomorphic recursively saturated models. This follows from a simple cardinality argument. Each countable model of T has an elementary extension to a countable recursively saturated model. If among all those models were less than $2^{\aleph_0}$ nonisomorphic ones, that would imply that there are less than $2^{\aleph_0}$ complete types of T. Still we have a very useful almost-uniqueness variant of Theorem 13.18 with the same proof.

THEOREM 13.33. *Let $\mathfrak{M}$ and $\mathfrak{N}$ be countable recursively saturated models of a complete theory. Then $\mathfrak{M}$ is isomorphic to $\mathfrak{N}$ if and only if $\mathfrak{M}$ and $\mathfrak{N}$ realize the same complete types.*

EXERCISE 13.34. Prove Theorem 13.33.

13.5.1. Types in Recursively Saturated Models of Arithmetic.
Recall that true arithmetic TA, is the complete theory of the standard model $(\mathbb{N}, +, \cdot)$. A model $\mathfrak{M}$ of TA is simple, if $M = \mathrm{Scl}(c)$, for some c in M.

EXERCISE 13.35. Show that simple models of TA are not recursively saturated.

Theorem 13.33 has an interesting variant for models of arithmetic. It will be formulated in terms of *standard systems*. In the following definition, $x|y$ is the divisibility relation defined in every model of TA by the formula $\exists z(x \cdot z = y)$.

DEFINITION 13.36. For $n \in \mathbb{N}$, let p_n be the n-th prime number, and let $\mathfrak{M}$ be a nonstandard model of TA. For $a \in M$, let D_a be $\{n : \mathfrak{M} \models p_n | a\}$. The *standard system* of $\mathfrak{M}$, denoted $\mathrm{SSy}(\mathfrak{M})$, is the set $\{D_a \cap \mathbb{N} : a \in M\}$.

Notice that the set D_a is a parametrically definable subset of M. In particular, by the overspill principle, if $D_a \cap \mathbb{N}$ is infinite, then it must contain some nonstandard elements of M.

EXERCISE 13.37. Use the compactness theorem to show that for every $A \subseteq \mathbb{N}$, there is a countable model $\mathfrak{M}$ os TA such that A is in $\mathrm{SSy}(\mathfrak{M})$.

THEOREM 13.38. *Let $\mathfrak{M}$ and $\mathfrak{N}$ be countable recursively saturated models of TA. Then, $\mathfrak{M}$ is isomorphic to $\mathfrak{N}$ if and only if $\mathrm{SSy}(\mathfrak{M}) = \mathrm{SSy}(\mathfrak{N})$.*

PROOF. Clearly, if $\mathfrak{M}$ is isomorphic to $\mathfrak{N}$, then $\mathrm{SSy}(\mathfrak{M}) = \mathrm{SSy}(\mathfrak{N})$.

Let us assume now that $\mathrm{SSy}(\mathfrak{M}) = \mathrm{SSy}(\mathfrak{N})$. By Theorem 13.33, we need to show that $\mathfrak{M}$ and $\mathfrak{N}$ realize the same complete types. Because in every model of TA finite sequences are coded by single elements, it will suffice to show that $\mathfrak{M}$ and $\mathfrak{N}$ realize the same complete 1-types.

Recall that for a formula φ, $\ulcorner\varphi\urcorner$ denotes its Gödel number (see Section 12.4).

Let $a \in M$ be given. Consider the type $q(x, a)$:

$$\{p_{\ulcorner\varphi(x)\urcorner}|x \iff \varphi(a) : \varphi \in \mathcal{L}_{\mathsf{TA}}\}$$

Notice that x occurs as a variable in $p_{\ulcorner\varphi(x)\urcorner}|x \iff \varphi(a)$ only once.

The set $\{\ulcorner\varphi\urcorner : \varphi \in \mathcal{L}_{\mathsf{TA}}\}$ is computable and so is the function $n \mapsto p_n$; hence $q(x, a)$ is computable.

Let $\varphi_0(x), \ldots, \varphi_n(x)$ be formulas of $\mathcal{L}_{\mathsf{TA}}$, and let

$$I = \{\ulcorner\varphi_i(x)\urcorner : i \leq n \text{ and } \mathfrak{M} \models \varphi_i(a)\}.$$

Then all the formulas $p_{\ulcorner\varphi_i(x)\urcorner}|x \iff \varphi_i(a)$ for $i \leq n$, are satisfied in the standard model by $\Pi_{i \in I} p_i$.

By recursive saturation, $q(x, a)$ is realized by some $d \in M$. Then $D_d \cap \mathbb{N} = \{\ulcorner \varphi(x) \urcorner : \mathfrak{M} \models \varphi(a)\}$ is in $\mathrm{SSy}(\mathfrak{M})$. In other words, the set of Gödel numbers of the formulas in $\mathrm{tp}(a)$ is in $\mathrm{SSy}(\mathfrak{M})$; hence, also in $\mathrm{SSy}(\mathfrak{N})$. Let $e \in N$ be such that $D_e \cap \mathbb{N} = D_d$, and consider the type $r(x, e)$

$$\{\varphi(x) \Longleftrightarrow p_{\ulcorner \varphi(x) \urcorner} | e : \varphi \in \mathcal{L}_{\mathsf{TA}}\}.$$

Since $\mathfrak{M}$ and $\mathfrak{N}$ are elementarily equivalent, $r(x, e)$ is finitely realizable in $\mathfrak{N}$; hence, by recursive saturation, $r(x, e)$ is realized in $\mathfrak{N}$. If b realizes $r(x, e)$, then $\mathrm{tp}^{\mathfrak{M}}(a) = \mathrm{tp}^{\mathfrak{N}}(b)$ and this finishes the proof. $\qquad \square$

13.5.2. Truth and Recursive Saturation. By Theorem 13.29, every consistent theory in a finite language has a countable recursively saturated model. We will finish this chapter with another proof of this result that involves elementary extensions of the standard model of arithmetic.

If $(\mathbb{N}, +, \cdot, \mathfrak{X})$ is an expansion of the standard model of arithmetic by a set $\mathfrak{X}$ of relations, functions, and constants, then the induction axioms

$$\forall \bar{y}\{[\varphi(0, \bar{y}) \wedge (\forall x \varphi(x, \bar{y}) \Longrightarrow \varphi(x + 1, \bar{y}))] \Longrightarrow \forall x \varphi(x, \bar{y})\},$$

hold for all formulas $\varphi(x, \bar{y})$ of the language of arithmetic extended by symbols for all new relations, functions, and constants, and it follows that the overspill principle (Proposition 12.9) holds for nonstandard models of $\mathrm{Th}((\mathbb{N}, +, \cdot, \mathfrak{X}))$.

For the main result in this section, we will need the following lemma.

LEMMA 13.39. *For every nonstandard model $\mathfrak{M}$ of* TA, *all arithmetic sets are in* $\mathrm{SSy}(\mathfrak{M})$.

PROOF. Let $A \subseteq \mathbb{N}$ be an arithmetic set defined in the standard model by $\varphi(x)$. Let $\mathfrak{M}$ be a nonstandard model of TA. The function $n \mapsto p_n$ is arithmetic, hence the relation $p_y | x$ is also arithmetic. For all $n \in \mathbb{N}$ we have

$$\mathfrak{M} \models \exists x \forall y [y < n \Longrightarrow (p_y | x \Longleftrightarrow \varphi(y))].$$

By the overspill principle, there is a nonstandard c such that

$$\mathfrak{M} \models \exists x \forall y [y < c \Longrightarrow (p_y | x \Longleftrightarrow \varphi(y))].$$

Let d be such that $\mathfrak{M} \models \forall y < c(p_y | d \Longleftrightarrow \varphi(y))$. Then $A = D_d \cap \mathbb{N} \in \mathrm{SSy}(\mathfrak{M})$. $\qquad \square$

As in Theorem 12.11, let $\mathcal{T} = \{(\ulcorner \varphi(x) \urcorner, n) : (\mathbb{N}, +, \cdot) \models \varphi(n)\}$.

PROPOSITION 13.40. *The reduct to the language of arithmetic of any elementary extension of $(\mathbb{N}, +, \cdot, \mathcal{T})$ is recursively saturated.*

PROOF. Let $\mathfrak{M}_{\mathcal{S}} = (M, +, \cdot, \mathcal{S})$ be an elementary extension of $(\mathbb{N}, +, \cdot, \mathcal{T})$, and let $\mathfrak{M} = (M, +, \cdot)$ be the reduct of $\mathfrak{M}_{\mathcal{S}}$ to the language of TA. We will show that $\mathfrak{M}$ is recursively saturated.

Since $\mathfrak{M}$ is a model of TA, it is enough to consider types with single parameters. Let $p(x, a)$, for some $a \in M^{<\omega}$, be a computable type that is finitely realizable in $\mathfrak{M}$.

For each $\varphi(x,y)$ let $\varphi^*(z)$ be $\varphi((z)_0, (z)_1)$. Let $P = \{\ulcorner\varphi^*(z)\urcorner : \varphi(x,a) \in p(x,a)\}$. It can be shown that the function $\ulcorner\varphi(x,y)\urcorner \mapsto \ulcorner\varphi^*(z)\urcorner$ is definable in the standard model. Hence P is arithmetic and, by Lemma 13.39, P is in $\mathrm{SSy}(\mathfrak{M})$. Then, $P = D_e \cap \mathbb{N}$ for some $e \in \mathfrak{M}$. Let S be the binary relation symbol whose interpretation in $\mathfrak{M}_S$ is $\mathcal{S}$. For each $n \in \mathbb{N}$ we have

$$\mathfrak{M}_S \models \exists x \forall y < n[y|e \Longrightarrow S(y, \langle x, a \rangle)],$$

where $(x, y) \mapsto \langle x, y \rangle$ is Cantor's pairing function. By the overspill principle,

$$\mathfrak{M}_S \models \exists x \forall y < c[y|e \Longrightarrow S(y, \langle x, a \rangle)],$$

for some nonstandard c. Then, if $b \in M$ is such that

$$\mathfrak{M}_S \models \forall y < c[y|e \Longrightarrow S(y, \langle b, a \rangle)],$$

then b realizes $p(x, \bar{a})$. $\square$

Proposition 13.40 shows a special way in which recursively saturated models of TA can be constructed. It can be generalized to models of arbitrary theories. If $\mathfrak{M}$ is a countable model of a theory T in a finite language, then we can identify the domain of $\mathfrak{M}$ with $\mathbb{N}$. In this setting, $\mathfrak{M}$ can be expanded to the structure $(\mathbb{N}, +, \cdot, \dots)$, where the dots indicate the relations, functions, and constants of $\mathfrak{M}$. We expand this structure further by adding the relation

$$\mathcal{T}_{\mathfrak{M}} = \{(\ulcorner\varphi(x)\urcorner, n) : \mathfrak{M} \models \varphi(n)\},$$

where $\varphi(x)$ ranges over all formulas of the language of $\mathfrak{M}$. Then the reduct of $(\mathbb{N}, +, \cdot, \dots, \mathcal{T}_{\mathfrak{M}})$ to the language of $\mathfrak{M}$ is recursively saturated.

EXERCISE 13.41. Prove the last statement above.

CHAPTER 14

Automorphisms of Recursively Saturated Structures

The notion of recursive saturation was introduced by Jon Barwise and John Schlipf in [**BS76**]. Countable recursively saturated structures are *resplendent*. Resplendence is a second-order notion of saturation. We will not define resplendence here, but it is this property that has turned out to be very useful in applications (see [**Kos11**]). Instead, we will see several results on automorphisms. Each countable recursively saturated structure has a rich automorphism group and one can learn much about those structures by studying their automorphism groups. This chapter is a short introduction to this area of research.

Let us begin with an example. Let T be $\mathrm{Th}((\mathbb{N}, <))$. Let us expand the language of T by adding a new unary relation symbol E, and let T_E be T together with the axiom:

$$[\forall y \neg \mathrm{Succ}(y, x) \implies E(x)] \wedge [\forall x, y(\mathrm{Succ}(x, y) \implies (E(x) \iff \neg E(y)))].$$

EXERCISE 14.1. Show that $(\mathbb{N}, <)$ has exactly one expansion to a model of T_E.

As in the proof of Theorem 8.4, let $(N, <)$ be an elementary extension of $(\mathbb{N}, <)$. As we have seen, $N \setminus \mathbb{N}$ is the union of isomorphic copies of $(\mathbb{Z}, <)$. Each copy of $(\mathbb{Z}, <)$ has its own set of even numbers, so let us define E_1 to be the set of all those even numbers, and let E be the set of even numbers of $\mathbb{N}$. Then $(N, <, E \cup E_1)$ is an expansion of $(N, <)$ to a model of T_E. The function $f : N \longrightarrow N$ that shifts all elements of $N \setminus \mathbb{N}$ up by 1 is an automorphism of $(N, <)$ and f maps E_1 onto the set $E_2 = N \setminus (\mathbb{N} \cup E_1)$. Then, because f is an automorphism, $(N, <, E \cup E_2)$ is also a model of T_E; hence $(N, <)$ has at least two expansions to models of T_E.

The argument above shows that the set of even numbers is not parametrically definable in $(\mathbb{N}, <)$ and we know it already, as it follows from Theorem 8.4. The argument has been brought up here to motivate the following definition.

DEFINITION 14.2. Let T be a theory in a language $\mathcal{L} \cup \{R\}$, where R is a relation symbol. We say that T defines R implicitly if every structure for $\mathcal{L}$ is expandable to at most one model of T.

The following early result in model theory is due to Evert Beth.

THEOREM 14.3. *If a theory T in a language $\mathcal{L} \cup \{R\}$ defines R implicitly, then there is a formula $\varphi(\bar{x})$ of $\mathcal{L}$ such that for every model $\mathfrak{M}$ of T, $\mathfrak{M} \models \forall \bar{x}[R(\bar{x}) \Longleftrightarrow \varphi(\bar{x})]$.*

A proof Beth's theorem that we will present involves automorphisms of recursively saturated structures and this is why it is included in this chapter. For a discussion of the theorem and two closely related results—Robinson's consistency theorem and Craig's interpolation theorem—see [**Kei77**] or [**Doe96**, Section 4.7].

For the rest of this section, we fix a countable recursively saturated structure $\mathfrak{M}$ for a finite language $\mathcal{L}$ and a computable enumeration $\{\varphi_n(x) : n \in \mathbb{N}\}$ of all $\mathcal{L}$-formulas with one free variable. Such an enumeration exists, because we have assumed that $\mathcal{L}$ is finite. Under this assumption there is an algorithmic procedure that generates all formulas of $\mathcal{L}$.

Recall that a is definable in $\mathfrak{M}$ if for some formula $\varphi(x)$

$$\mathfrak{M} \models [\varphi(a) \wedge \exists! x \varphi(x)].$$

PROPOSITION 14.4. *For a in M, the following are equivalent:*

(1) *a is definable in $\mathfrak{M}$.*
(2) *The only element of M that realizes $\mathrm{tp}(a)$ is a.*

PROOF. (1) $\Longrightarrow$ (2) is obvious. For the proof of (2) $\Longrightarrow$ (1), assume (2) and consider the type

$$p(x,a) = \{ \bigwedge_{i \le n} [\varphi_i(x) \Longleftrightarrow \varphi_i(a)] : n \in \mathbb{N} \} \cup \{a \ne x\}.$$

Since a is the only element of M realizing $\mathrm{tp}(a)$, by recursive saturation, $p(x,a)$ is not finitely realizable in $\mathfrak{M}$. Hence there is n such that

$$\mathfrak{M} \models \forall x[\bigwedge_{i \le n} [\varphi_i(x) \Longleftrightarrow \varphi_i(a)] \Longrightarrow x = a]. \tag{$*$}$$

Fix an n as above, and let

$$\theta(x) = \bigwedge \{\varphi_i(x) : i \le n \wedge \mathfrak{M} \models \varphi_i(a)\} \wedge \bigwedge \{\neg\varphi_j(x) : j \le n \wedge \mathfrak{M} \models \neg\varphi_j(a)\}.$$

Then it follows from $(*)$ that

$$\mathfrak{M} \models \theta(a) \wedge \exists! x \theta(x).$$

Thus, a is definable in $\mathfrak{M}$. □

Because countable recursively saturated models are strongly homogeneous (Corollary 13.31), we have the following corollary.

COROLLARY 14.5. *For a in M, the following are equivalent:*

(1) *a is definable in $\mathfrak{M}$.*
(2) *For all f in $\mathrm{Aut}(\mathfrak{M})$, $f(a) = a$.*

PROPOSITION 14.6. *For each $\bar{a}$ in $M^{<\omega}$, there is a b in M that is not definable in $(\mathfrak{M}, \bar{a})$.*

PROOF. For $\bar{a}$ in $M^{<\omega}$ consider the type.

$$p(x, \bar{a}) = \{\exists! y \varphi(y, \bar{a}) \implies \neg\varphi(x, \bar{a}) : \varphi(x, \bar{y}) \in \mathcal{L}\}.$$

Since M is infinite, $p(x, \bar{a})$ is finitely realizable in $\mathfrak{M}$; hence it is realized. No b realizing $p(x, \bar{a})$ in $\mathfrak{M}$ is definable in $(\mathfrak{M}, \bar{a})$. □

By Propositions 14.4 and 14.6, there are an a in M and an automorphism f of $\mathfrak{M}$ such that $f(a) \neq a$. The following result is stronger and it will be crucial in the proof of Beth's theorem.

PROPOSITION 14.7. If $A \subseteq M^n$, for some $n > 0$, is undefinable and the expanded structure $(\mathfrak{M}, A)$ is recursively saturated, then there is an $f \in \text{Aut}(\mathfrak{M})$ such that $f(A) \neq A$.

PROOF. Let A be as in the proposition. We will use A for the relation symbol for the set A. Because $\mathfrak{M}$ is strongly homogeneous, it is enough to prove that there are $\bar{a}$ in A and $\bar{b}$ not in A such that $\text{tp}^{\mathfrak{M}}(\bar{a}) = \text{tp}^{\mathfrak{M}}(\bar{b})$.

Let n be the arity of A. For $\bar{a}$ in M^n and each $k > 0$, let $\text{tp}^k(\bar{a})$ be the following k-th approximation to $\text{tp}^{\mathfrak{M}}(\bar{a})$.

$$\{\theta_i(\bar{x}) : i \leq k \text{ and } \mathfrak{M} \models \theta_i(\bar{a})\} \cup \{\neg\theta_j(\bar{x}) : j \leq k \text{ and } \mathfrak{M} \models \neg\theta_j(\bar{a})\},$$

where $\{\theta_k(\bar{x}) : k \in \mathbb{N}\}$ is a computable enumeration of all $\mathcal{L}$-formulas with n free variables.

For each k, let E_k be defined by

$$(\bar{x}, \bar{y}) \in E_k \iff \text{tp}^k(\bar{x}) = \text{tp}^k(\bar{y}).$$

For each k, E_k is a definable equivalence relation on M^n. Because there are at most 2^{k+1} possible sequences of truth values assigned to the $\theta_0(\bar{a}), \ldots \theta_k(\bar{a})$, E_k has at most 2^{k+1} equivalence classes.

Now it is time to use recursive saturation of $(\mathfrak{M}, A)$. Consider the type $p(\bar{x}, \bar{y})$

$$\{A(\bar{x}) \wedge \neg A(\bar{y})\} \cup \{E_k(\bar{x}, \bar{y}) : k \in \mathbb{N}\}.$$

Because $p(\bar{x}, \bar{y})$ is computable, the proof will be finished if we show that this type is finitely realizable in $(\mathfrak{M}, A)$. So suppose it is not. Then, for some k

$$\mathfrak{M} \models \forall \bar{x}, \bar{y}[E_k(\bar{x}, \bar{y}) \implies (A(\bar{x}) \iff A(\bar{y}))].$$

Then each equivalence class of E_k is either disjoint from A or is contained in A. Let $\bar{a}_1, \ldots, \bar{a}_l$ be the set of representatives for all equivalence classes of E_k that are contained in A. And let $\theta_i(\bar{x})$ be the conjunction of all formulas in $\text{tp}^k(\bar{a}_i)$, for $i \leq l$. It is easy to check that for all $\bar{a}$ in M^n

$$\bar{a} \in A \iff \mathfrak{M} \models [\theta_0(\bar{a}) \vee \cdots \vee \theta_l(\bar{a})].$$

So, contrary to the assumption, A is definable in $\mathfrak{M}$, which finishes the proof. □

14.0.1. Proof of Beth's theorem. Suppose that a theory T in a language $\mathcal{L} \cup \{R\}$ defines R implicitly. Without loss of generality, we can assume that the $\mathcal{L}$ is finite.

EXERCISE 14.8. Prove the last statement above. HINT: Use the compactness theorem to show that if a theory T in a language $\mathcal{L} \cup \{R\}$ defines R implicitly, then there is a finite $T' \subseteq T$ that defines R implicitly.

If T is inconsistent, then there is nothing to prove. So let us assume that T is consistent. We need to show that there is an $\mathcal{L}$-formula $\varphi(\bar{x})$ such for any model $(\mathfrak{M}, R)$ of T, $\varphi(\bar{x})$ defines R in $\mathfrak{M}$.

Let n be the arity ot R. As before, let $\{\theta_k(\bar{x}) : k \in \mathbb{N}\}$ be a computable enumeration of all formulas of $\mathcal{L}$ with n free variables. If there is no formula that defines R is all models of T, then the following theory T' is finitely consistent

$$T \cup \{\exists \bar{x}[\neg R(\bar{x}) \Longleftrightarrow \theta_k(\bar{x})] : k \in \mathbb{N}\}.$$

Let $(\mathfrak{M}, R)$ be a model of T'. Since $\mathcal{L}$ is finite, we can assume that M is countable and by Proposition 13.29 we can further assume that $(\mathfrak{M}, R)$ is recursively saturated. By Proposition 14.7, there is a automorphism f of $\mathfrak{M}$ such that $f(R) \neq R$. Then $(\mathfrak{M}, R)$ and $(\mathfrak{M}, f(R))$ are distinct expansions of $\mathfrak{M}$ to a model of T. Contradiction.

14.1. Large Automorphism Groups

$\mathfrak{M}$ is still a countable recursively saturated model of a theory in a finite language $\mathcal{L}$.

We will use the following two propositions to show that any countable recursively saturated structure has lots and lots of automorphisms.

PROPOSITION 14.9. Let $A \subseteq M$ be parametrically definable and infinite. Then, there are distinct a and b in A such that $\text{tp}(a) = \text{tp}(b)$.

PROOF. Recall that $\{\varphi_n(x) : n \in \mathbb{N}\}$ is a computable enumeration of all formulas of the language of $\mathfrak{M}$ with one free variable. For $k \in \mathbb{N}$, let $\theta_k(x, y)$ be the formula $\bigwedge_{i \leq k}(\varphi_i(x) \Longleftrightarrow \varphi_i(y))$.

For each k, $\theta_k(x, y)$ defines an equivalence relation R_k on M. Because for each a in M, there are only 2^{k+1} possible assignments of truth values for $\varphi_0(a), \ldots, \varphi_k(a)$, for each k, R_k has finitely many equivalence classes.

Let $\psi(x, \bar{c})$, be a formula defining A in $\mathfrak{M}$, for some $\bar{c}$ in $M^{<\omega}$. Consider the 2-type

$$p(x, y, \bar{c}) = \{(x \neq y \wedge \psi(x, \bar{c}) \wedge \psi(y, \bar{c})\} \cup \{\theta_k(x, y) : k \in \mathbb{N}\}.$$

Because $\{\theta_k(x, y) : k \in \mathbb{N}\}$ is a computable sequence of formulas, $p(x, y, \bar{c})$ is a computable type. Hence, if $p(x, \bar{c})$ is finitely realizable in $\mathfrak{M}$, then it is realized by some a and b in A such that $\text{tp}(a) = \text{tp}(b)$, so we are done. To get a contradiction, let us assume that $p(x, y, \bar{c})$ not finitely realizable in $\mathfrak{M}$. Then there is a k such that

$$\mathfrak{M} \models \forall x, y[(\psi(x, \bar{c}) \wedge \psi(y, \bar{c}) \wedge \theta_k(x, y, \bar{c}))] \Longrightarrow x = y,$$

implying that each equivalence class of R_k has at most one element. Because R_k has finitely many equivalence classes and A is infinite, we get a contradiction. $\square$

We get a direct corollary:

COROLLARY 14.10. Let $A \subseteq M$ be parametrically definable and infinite. Then, there is an a and automorphism f of $\mathfrak{M}$ such that $a \neq f(a) \in A$.

By Corollary 14.10, if $\mathfrak{M}$ is a countable nonstandard model of TA, then for every a and b in M such that $a + n < b$, for all standard n, there are a c such that $a < c < b$ and an automorphism f of $\mathfrak{M}$ such that $f(c) \neq c$.

The next corollary shows even more and it applies to all countable recursively saturated ordered structures.

COROLLARY 14.11. If $\mathfrak{M}$ is a countable recursively saturated ordered structure and $A \subseteq M$ is parametrically definable and infinite, then there is an infinite $B \subseteq A$ such that for all a and b in B, $\mathrm{tp}(a) = \mathrm{tp}(b)$.

PROOF. Suppose that $\psi(x, \bar{a})$ defines A in $\mathfrak{M}$. Then, $\psi(x, \bar{a})$ defines A in $(\mathfrak{M}, \bar{a})$ without parameters. Since $(M, \bar{a})$ is recursively saturated, by Corollary 14.10 there is an automorphism f of $(\mathfrak{M}, \bar{a})$ such that $a \neq f(a)$. Moreover, since f fixes $\bar{a}$, for all b in A, $f(b)$ is in A. We can assume that $a < f(a)$, because otherwise we can take the inverse f^{-1} instead of f. Let $a_0 = a$, and for all n, let $a_{n+1} = f(a_n)$, and let $B = \{a_n : n \in \mathbb{N}\}$. By the previous remark, $B \subseteq A$. Because f preserves the ordering of $\mathfrak{M}$, $a_n < a_{n+1}$, for all n; hence B is infinite. $\square$

For Proposition 14.9 the assumption that $\mathfrak{M}$ is countable is not needed. It is needed for the following theorem. It shows that countable recursively saturated structures have as many automorphisms as possible. To prove it, we will modify the back-and-forth argument used in the proof of Theorem 9.11. In that proof, countability and homogeneity of $\mathfrak{M}$ is used to show that if for $\bar{b}$ and $\bar{c}$ in $M^{<\omega}$ if $\mathrm{tp}(\bar{b}) = \mathrm{tp}(\bar{c})$, then there is an automorphism f such that $f(\bar{b}) = \bar{c}$. Here we will apply Proposition 14.4 to show that such an f can be constructed in many different ways, giving us the result. We will use a construction involving the set of all finite 0-1 sequences, denoted $2^{<\omega}$. Let 2^ω be the set of all infinite 0-1 sequences. For a sequence α by $\alpha \restriction n$ we will denote the restriction of α to $\{0, \ldots, n\}$. In the construction, for each sequence $\sigma \in 2^{<\omega}$, we will define a one-to-one function f_σ between two finite subsets of M in such that

- For σ and τ in $2^{<\omega}$, and n in $\mathbb{N}$, if $\tau = \sigma \restriction n$, then $f_\tau \subseteq f_\sigma$.
- For each $\alpha \in 2^\omega$, the function $f_\alpha = \bigcup_{n \in \mathbb{N}} f_{\alpha \restriction n}$ is in $\mathrm{Aut}(\mathfrak{M})$.
- For each $\sigma \in 2^{<\omega}$ there are $\tau_0, \tau_1 \in 2^{<\omega}$ extending σ such that $f_{\tau_0}(a) \neq f_{\tau_1}(a)$ for some a in the domains of both functions.

By the conditions above, for distinct α and β in 2^ω, $f_\alpha \neq f_\beta$, and since 2^ω is of cardinality $2^{\aleph_0}$, this will give us $2^{\aleph_0}$ automorphisms of $\mathfrak{M}$.

THEOREM 14.12. $|\operatorname{Aut}(\mathfrak{M})| = 2^{\aleph_0}$.

PROOF. For each σ in $2^{<\omega}$ we will define two sequences $\bar{b}_\sigma$ and $\bar{c}_\sigma$ in $M^{<\omega}$, such that $\operatorname{tp}(\bar{b}_\sigma) = \operatorname{tp}(\bar{c}_\sigma)$ and if σ extends τ, then $\bar{b}_\sigma$ extends $\bar{b}_\tau$ and $\bar{c}_\sigma$ extends $\bar{c}_\tau$.

Let $\{a_n : n \in \mathbb{N}\}$ be an enumeration of M, and let $b_\varnothing = c_\varnothing = a_0$.

Assume now that $\bar{b}_\sigma$ and $\bar{c}_\sigma$ have been defined.

By Propositions 14.4 and 14.6, there are elements of M that are not definable in $(\mathfrak{M}, \bar{b}_\sigma)$. Let b be the first a_i that is such an element. Let $\bar{b}_{\sigma 0} = \bar{b}_{\sigma 1} = \bar{b}_\sigma, b$. Then, using Proposition 14.4 and homogeneity of $\mathfrak{M}$, we find distinct c and c' such that $\operatorname{tp}(\bar{b}_\sigma, b) = \operatorname{tp}(\bar{c}_\sigma, c) = \operatorname{tp}(\bar{c}_\sigma, c')$, and we let $\bar{c}_{\sigma 0} = \bar{c}_\sigma, c$ and $\bar{c}_{\sigma 1} = \bar{c}_\sigma, c'$.

After that, we perform two back-and-forth steps as in the proof of Theorem 9.13, extending the sequences to guarantee that every element of M eventually gets included in the domains and ranges of the automorphisms we are constructing.

The construction guarantees that for every α in 2^ω, f_α that is the union of all partial maps $\bar{b}_{\alpha \restriction n} \mapsto \bar{c}_{\alpha \restriction n}$, is an automorphism of $\mathfrak{M}$. If $\alpha \neq \beta$ then $f_\alpha \neq f_\beta$ and this finishes the proof. $\square$

14.1.1. Automorphisms of Models of Arithmetic. The standard model of arithmetic is rigid and its simple elementary extensions are rigid as well (see the remark after Proposition 12.6). In contrast, the automorphism groups of countable recursively saturated models of TA and PA, are not only of size $2^{\aleph_0}$, but they have complex group-theoretic structure that has been a subject of interesting research that involves a mixture of model-theoretic techniques and group theory. Here we will just mention some results related to the material covered in these lectures. For simplicity, they will be formulated just for models of TA, with small modifications they hold for all models of PA. For a full account see [**KM94**] and [**KS06**].

Let us begin with a simple observation. Let $\mathfrak{M}$ be a countable recursively saturated model of TA. It is easy to see that no nontrivial automorphism of $\mathfrak{M}$ is definable. To see this, suppose that an automorphism f is defined in $\mathfrak{M}$ by a formula $\varphi(x, y)$, and let $\psi(x)$ be $\varphi(x, x)$. If for some $a \in M$, $f(a) = a$, then $f(a + 1) = f(a) + f(1) = a + 1$; hence we have

$$M \models \psi(0) \wedge \forall x[\psi(x) \implies \psi(x + 1)].$$

It follows that $\mathfrak{M} \models \forall x \psi(x)$, so f is identity.

For a detailed account of Presburger arithmetic Pr see [**Smo91**]. The results discussed there combined with the material from Section 13.5.2 can be used to show that for every nonstandard model $\mathfrak{M}$ of TA, the reduct $(M, +)$ is a recursively saturated model of Pr and that every countable recursively saturated model of Pr can be expanded to a model of TA, moreover, there are $2^{\aleph_0}$ such expansions that are pairwise nonisomorphic. Among those expansions there are $2^{\aleph_0}$ recursively saturated models of TA, but any two such expansions are isomorphic.

Countable recursively saturated structure have $2^{\aleph_0}$ automorphisms. A definable subset of a structure cannot be moved by any automorphism and a parametrically definable subset can have at most $\aleph_0$ automorphic images, and that is because the image a set defined by $\varphi(x, \bar{a})$ under an automorphism f is defined by $\varphi(x, f(\bar{a}))$. There are extensions of first-order logic in which some first-order undefinable sets can be defined. Such is for example the set of all elements of the model realizing a given complete type. Such sets cannot be moved by automorphisms, and they can be defined in the infinitary logic known as $L_{\omega_1, \omega}$ in which infinite conjunctions and disjunctions are allowed. If a set has uncountably many automorphic images, this shows that the set cannot be definable in such extensions. Examples of such sets are given below.

All results discussed in the rest of this section can be found in [**KS06**].

As in Section 13.5.2, let $\mathcal{T} = \{(\ulcorner\varphi(x)\urcorner, n) : (\mathbb{N}, +, \cdot) \models \varphi(n)\}$. Proposition 13.40 states that if $(M, +, \cdot, \mathcal{S})$ is an elementary extension of $(\mathbb{N}, +, \cdot, \mathcal{T})$, then the reduct $(M, +, \cdot)$ is recursively saturated. For countable models there is a converse: if $\mathfrak{M}$ is a countable recursively saturated model of TA, then $\mathfrak{M}$ can be expanded to $(\mathfrak{M}, \mathcal{S})$ that is an elementary extension of $(\mathbb{N}, +, \cdot, \mathcal{T})$. It can be shown that if $(\mathfrak{M}, \mathcal{S})$ is such an extension, then $\mathcal{S}$ has $2^{\aleph_0}$ automorphic images, i.e., the set $\{f(\mathcal{S}) : f \in \mathrm{Aut}(\mathfrak{M})\}$ is of cardinality $2^{\aleph_0}$.

The conjugacy class of an element f in a group G is $\{gfg^{-1} : g \in G\}$. The conjugacy class of every nontrivial automorphism of a countable recursively saturated model of TA is of cardinality $2^{\aleph_0}$. Because the graph of gfg^{-1} is the automorphic image of the graph of f, it follows that the graph of each nontrivial automorphism of a countable recursively saturated model of TA has $2^{\aleph_0}$ automorphic images.

There are uncountable recursively saturated models of TA that are rigid and there are uncountable recursively saturated models of TA that cannot be expanded to an elementary extension of $(\mathbb{N}, +, \cdot, \mathcal{T})$.

Let $\mathfrak{M}$ be countable recursively saturated model of TA. Recall that the standard system of $\mathfrak{M}$, $\mathrm{SSy}(\mathfrak{M})$ is the set of intersections with $\mathbb{N}$ of all the subsets of M that are coded by elements of M (Definition 13.36). By Corollary 14.5, any nonstandard element of M can be moved by an automorphism. One can ask if there is an automorphism that moves all nonstandard elements. It turns out that there are such automorphisms if and only if $\mathrm{SSy}(\mathfrak{M})$ is closed under arithmetic definability, which means that for every A in $\mathrm{SSy}(\mathfrak{M})$, if B is definable in $(\mathfrak{M}, A)$, then B is in $\mathrm{SSy}(\mathfrak{M})$. We call models of TA satisfying this condition *arithmetically saturated*. If $\mathfrak{N}$ is a countable arithmetically saturated model of TA, then $\mathfrak{N}$ is isomorphic to $\mathfrak{M}$ if and only if $\mathrm{Aut}(\mathfrak{N})$ is isomorphic to $\mathrm{Aut}(\mathfrak{M})$. This and related results show how complex automorphism groups of recursively saturated and arithmetically saturated models of arithmetic are.

14.2. Automorphisms of Complex Numbers

As we have seen in Chapter 5, the field of real numbers is rigid. Every rational number is definable, and every irrational number is tightly squeezed

between the rational numbers below and above it, so it cannot be moved by
any automorphism. In contrast, the field of complex numbers $\mathfrak{C} = (\mathbb{C}, +, \cdot)$
is of cardinality $2^{\aleph_0}$ and has the maximum number $2^{2^{\aleph_0}}$ of automorphisms.
This follows from two theorems in algebra. The first is that $\mathfrak{C}$ has a transcendence basis of cardinality $2^{\aleph_0}$, and the second that every permutation of a
transcendence basis extends to an automorphism of $\mathfrak{C}$. However, no nontrivial
automorphism of $\mathfrak{C}$ is definable in $\mathfrak{C}$.

One could object to the last claim above by invoking conjugation, i.e the
function $\alpha(a + bi) = a - bi$. It is easy to see that conjugation is indeed an
automorphism and it seems that it is definable, after all, I just wrote a formula
for it. However, if conjugation were definable in $\mathfrak{C}$, then so would be the set of
its fixed points $\text{fix}(\alpha) = \{z : \alpha(z) = z\}$, but $a + bi = a - bi$ if only if $b = 0$,
hence $\text{fix}(\alpha) = \mathbb{R}$ and this is impossible. As mentioned in Chapter 8, $\mathfrak{C}$ is a
minimal structure. All of its parametrically definable sets are either finite of
cofinite and $\mathbb{R}$ is definitely not such a set. What is going on here?

The question here is: what is $\mathfrak{C}$? $\mathfrak{C}$ can be defined as the unique algebraically closed field of cardinality $2^{\aleph_0}$ and characteristic 0. A familiar representation of $\mathfrak{C}$ is defined in the field $\mathfrak{R} = (\mathbb{R}, +, \cdot)$ as the set of all pairs (a, b),
with i being $(0, 1)$ and suitably defined addition and multiplication. The def-
initions are first-order, and, under this representation, α is definable in $\mathfrak{R}$ as
well, but as the argument above shows, it is not definable in $\mathfrak{C}$.

We will finish this chapter with a proof that shows the existence of a
nontrivial automorphism of $\mathfrak{C}$ that is not α. It is included here as another
illustration of the techniques discussed in this book.

For the rest of this section, let $\mathfrak{R}^*$ be an elementary extension of $\mathfrak{R}$.

EXERCISE 14.13. Show that $\mathfrak{R}^*$ is not isomorphic to $\mathfrak{R}$. HINT: Show that
$\mathfrak{R}^*$ must have an element c such that $r < c$, for all r in $\mathbb{R}$.

Let $\mathbb{R}^*$ be the domain of $\mathfrak{R}^*$. By the downward Löwenheim-Skolem Theo-
rem, we can assume that $|\mathbb{R}^*| = 2^{\aleph_0}$. Let $\mathbb{C}^* = (\mathbb{R}^*)^2$, and let $\mathfrak{C}^* = (\mathbb{C}^*, +, \cdot)$,
where $+$ and $\cdot$ are defined as in the representation of $\mathfrak{C}$ in $\mathfrak{R}$ mentioned above.
We will show that $\mathfrak{C}^*$ is algebraically closed.

Under the representation of $\mathfrak{C}$ in $\mathfrak{R}$, the fact that $\mathfrak{C}$ is an algebraically
closed field is expressed by a set of first-order sentences of the language of $\mathfrak{R}$,
comprised of the field axioms and an infinite set of sentences expressing that
all polynomial equations of degree n have solutions—one sentence for each n.
Because $\mathfrak{R}^*$ is an elementary extension of $\mathfrak{R}$, all those sentences are true in
$\mathfrak{R}^*$. If $\mathfrak{C}^*$ is defined in $\mathfrak{R}^*$, by the same formulas that define $\mathfrak{C}$ in $\mathfrak{R}$, then $\mathfrak{C}^*$ is
algebraically closed.

Because, $|\mathbb{C}^*| = 2^{\aleph_0}$, $\mathfrak{C}^*$ is isomorphic to $\mathfrak{C}$. Let $F : \mathbb{C}^* \longrightarrow \mathbb{C}$ be an
isomorphism. Let α^* be conjugation in $\mathfrak{C}^*$ and let $\beta : \mathbb{C} \longrightarrow \mathbb{C}$ be $F \circ \alpha^* \circ F^{-1}$.

EXERCISE 14.14. Show that β is an automorphism of $\mathfrak{C}$.

The set of fixed points of α^* is $\mathbb{R}^*$. Because the graph of β is an isomorphic
image under F of the graph of α^*, the set of fixed points of β is isomorphic to

$\mathfrak{R}^*$. The set of fixed points of α is $\mathbb{R}$ and $\mathfrak{R}^*$ is not isomorphic to $\mathfrak{R}$; hence $\beta \neq \alpha$.

CHAPTER 15

Stability

The task of model theory is twofold. Locally, for a given theory T, we want to know if models of T can be classified according to mathematically meaningful criteria and, globally, we want to classify all first order theories with respect to classifiability of their models. For example, as we have seen in these lectures, there are countably many countable models of $\text{Th}((\mathbb{N}, <))$ and their isomorphism types can be easily described; and there are $2^{\aleph_0}$ countable models of $\text{Th}((\mathbb{N}, +))$ and we did not even try to classify these models. One of the reasons for such a difference is that the former theory has only countably many complete types, while the latter has $2^{\aleph_0}$ of them. This observation is just a beginning.

The *main gap theorem*, due to Saharon Shelah, roughly says that for any first-order theory T in a countable language, either models of T are completely unclassifiable, or there is a system of isomorphism invariants for the models of T. Both disjuncts of this dichotomy are given precise formulations. To prove his result, Shelah developed a a classification theory based on a variety of subtle model-theoretic techniques. This material cannot be covered in introductory lectures. See [**CK90**, Section 7.1] or [**Mar02**, Chapter 6], or for the full account [**She90**]. Here we will just discuss the concept of $\aleph_0$-stability that will give us a chance to revisit some of the main themes discussed in these lectures. One of those themes was counting the number of complete types of a given theory and that included theories of the form $\text{Th}(\mathfrak{M}, a)_{a \in A}$, for $A \subseteq M$. Some of the most important results in Shelah's classification theory are based on calculation of such numbers.

We will begin with a specific result based on the material developed in Chapters 10 and 11.

15.1. Models Realizing Few Types

There are many uses of models generated by indiscernibles. Theorem 15.4 below serves as an example and it will also be used to prove Theorem 15.9 below. For some motivation now, compare the statement of the theorem with the following two observations about models of DLO.

EXAMPLE 15.1. A proper subset $D \subseteq \mathbb{Q}$ is a *Dedekind cut* if D has no largest element and for all p and q if $p < q$ and q is in D, then p is in D. For each Dedekind cut D, let $p_D(x)$ be the type

$$\{p < x : p \in D\} \cup \{x < q : q \in \mathbb{Q} \setminus D\}.$$

For each Dedekind cut D, $p_D(x)$ is is realized in $(\mathbb{R}, <, p)_{p \in \mathbb{Q}}$ and if D and E are different Dedekind cuts, then the realizations of $p_D(x)$ and $p_E(x)$ are different. Because there are $2^{\aleph_0}$ Dedekind cuts, the ordered structure $(\mathbb{R}, < , p)_{p \in \mathbb{Q}}$ realizes $2^{\aleph_0}$ complete 1-types in the language $\{<\}$ with constants for all rational numbers. So $(\mathbb{R}, <, p)_{p \in \mathbb{Q}}$ is a model of DLO of size $2^{\aleph_0}$ that realizes $2^{\aleph_0}$ types with parameters from a countable subset of the domain.

For an example of a large model of DLO that realizes fewer types, consider $(D, <)$ that is the union of a chain of ordered sets $\{(D_\alpha, <)\}_{\alpha < \aleph_1}$, such that for each α, $(D_\alpha, <)$ is isomorphic to $(\mathbb{Q}, <)$ and for all $\alpha < \beta$ each element of $D_\beta \setminus D_\alpha$ is larger that each element of D_α. Using the notation from Section 13.2, $(D, <)$ can be defined as $(\aleph_1, <) \otimes (\mathbb{Q}, <)$.

$(D, <)$ is a model of DLO and for each countable $B \subseteq D$ there is an α such that $B \subseteq D_\alpha$.

For a and b in $D \setminus D_\alpha$ let β be such that a and b are in D_β, and let $D' = D_\beta \setminus D_\alpha$. Then $(D', <)$ is a countable model of DLO; hence there is an automorphism f of $(D', <)$ such that $f(a) = b$. Moreover, f can be extended to an automorphism of $(D, <)$ that is the identity function on $D \setminus D'$. It follows that $(D, <, b)_{b \in B}$ realizes only $\aleph_0$ complete 1-types; there are $\aleph_0$ such types realized in $(D_\alpha, <, b)_{b \in B}$ and one more realized by each element above D_α.

In preparation for the next result, let us recall some material from Chapter 11.

In the following two exercises, $\mathfrak{M}$ is a model for a theory with built-in Skolem functions and $\mathcal{L}$ is the language of $\mathfrak{M}$.

EXERCISE 15.2. Suppose that $A = \{a_i : i \in I\}$ is an indiscernible sequence in $\mathfrak{M}$ for an ordered set $(I, <)$. Show that for every n-ary Skolem function f and all increasing n-tuples $\bar{a}$ and $\bar{b}$ in A^n, $\mathrm{tp}^{\mathfrak{M}}(f(\bar{a})) = \mathrm{tp}^{\mathfrak{M}}(f(\bar{b}))$.

EXERCISE 15.3. For $B \subseteq C \subseteq M$, let $\mathfrak{M}(B)$ be the expansion $(\mathfrak{M}, b)_{b \in B}$. Show that $\mathrm{Scl}^{\mathfrak{M}}(C) = \mathrm{Scl}^{\mathfrak{M}(B)}(C \setminus B)$. HINT: It follows directly from definitions, but write down a complete argument.

THEOREM 15.4. Let $\mathcal{L}$ be a countable language, and let T be an $\mathcal{L}$-theory with infinite models. Then for every cardinal κ, T has a model $\mathfrak{M}$ such that $|M| = \kappa$ and for every countable $B \subseteq M$, the expanded model $\mathfrak{M}(B)$ realizes at most $\aleph_0$ complete 1-types.

PROOF. Because every type in a language $\mathcal{L}$ contains as subtypes all sets of formulas in sublanguages of $\mathcal{L}$ and our goal is to build a model with a small number of types, we may assume that T has built-in Skolem functions.

Let κ be an infinite cardinal, and let $(\kappa, <)$ be the ordered set of ordinal numbers smaller than κ with the usual ordering. Let $\mathfrak{M}$ be a model of T such that for some sequence of indiscernibles $A = \{a_\alpha : \alpha < \kappa\}$, $\mathrm{Sk}(\mathfrak{M}) = \mathrm{Scl}(A)$.

Such a model exists by Theorem 10.3. Because $\mathcal{L}$ is countable, $|M| = \kappa$. We will show that $\mathfrak{M}$ has the required property.

Let B be a countable subset of M. For each $b \in B$ pick an $\bar{a} \in A^{<\omega}$ and a Skolem function f such that $b = f(\bar{a})$. Let C be the set of all $a \in A$ that are terms in the sequences $\bar{a}$ chosen this way. Then, $B \subseteq \mathrm{Scl}(C)$ and $|C| \leq \aleph_0$. Let $\mathcal{L}(C)$ be the language of $\mathfrak{M}(C)$

We will use the ordering of A defined by: $a_\alpha < a_\beta$ if and only if $\alpha < \beta$.

By the definition of type, for $a \in M$,

$$\mathrm{tp}^{\mathfrak{M}(B)}(a) = \{\varphi(x, \bar{b}) : \bar{b} \in B^{<\omega} \text{ and } \mathfrak{M} \models \varphi(a, \bar{b})\}.$$

Our goal is to count the number of such types. Because every b in B is of the form $f(\bar{c})$ for some Skolem function f and $\bar{c}$ in $C^{<\omega}$, it follows that for all a and a' in M, $\mathrm{tp}^{\mathfrak{M}(C)}(a) = \mathrm{tp}^{\mathfrak{M}(C)}(a')$ if and only if $\mathrm{tp}^{\mathfrak{M}(B)}(a) = \mathrm{tp}^{\mathfrak{M}(B)}(a')$, so it is enough to count the number of complete 1-types of $\mathcal{L}(C)$ realized in $\mathfrak{M}$.

Suppose that for a Skolem function f, $d_1 = f(a_1, c_1)$ and $d_2 = f(a_2, d_2)$, where a_1, a_2 are in $A \setminus C$ and c_1, c_2 are in C. If follows from indiscernibility of A that $\mathrm{tp}^{\mathfrak{M}}(d_1) = \mathrm{tp}^{\mathfrak{M}}(d_2)$, whenever $a_1 < c_1$ iff $a_2 < c_2$. Because $\mathcal{L}$ is countable, there are only $\aleph_0$ Skolem functions, there are at most $\aleph_0$ complete types of elements of M of the form above.

In general, each element d of M is of the form $f(\bar{a}, \bar{c})$, for some Skolem function f, $\bar{a}$ in $(A \setminus C)^{<\omega}$ and $\bar{c}$ in $C^{<\omega}$. As in the simplified case of single a and c above, $\mathrm{tp}^{\mathfrak{M}}(d)$ is determined by f and the position of the terms of $\bar{a}$ among the terms of $\bar{c}$. Because there are only finitely many ways in which the terms of $\bar{a}$ can be placed among the terms of $\bar{c}$. This shows that there are at most $\aleph_0$ 1-types of elements of the form $f(\bar{a}, \bar{c})$ and finishes the proof. $\square$

EXERCISE 15.5. Fill in the details of the argument in the last paragraph of the proof above.

The proof of Theorem 15.4 follows the lines of the proof of a more general result in which there is no restriction on the cardinality of B and in the conclusion we get the expanded model $\mathfrak{M}(B)$ that realizes at most $|B| + \aleph_0$ types in the language $\mathcal{L}(B)$. See [**CK90**, Corollary 3.3.14].

15.2. $\aleph_0$-stable theories

Throughout this section, we assume that all languages are either finite or countable.

DEFINITION 15.6. A theory T is $\aleph_0$-stable if for every model $\mathfrak{M}$ of T and all countable $A \subseteq M$, $(\mathfrak{M}, a)_{a \in A}$ realizes at most $\aleph_0$ complete 1-types.

In Example 15.1, the example $(\mathbb{R}, <, p)_{p \in \mathbb{Q}}$ shows that DLO is not $\aleph_0$-stable.

As noted in the last remark in Section 9.5.2, DLO has $2^{\aleph_1}$ pairwise nonisomorphic models of cardinality $\aleph_1$. This is the maximum number of nonisomorphic models of this size. The proof does not reveal any way of classifying all those models. The diversity is forced by the sheer complexity of the ordered set of countable ordinals $(\aleph_1, <)$. The proof uses a partition of $\aleph_1$ into $\aleph_1$ pairwise disjoint stationary sets of countable ordinal numbers.

It turns out that models of all theories that admit infinitely ordered sets in the sense of the following definition are unclassifiable.

DEFINITION 15.7. A theory T has the *order property* if there are a model $\mathfrak{M} \models T$, sequences of tuples $\{\bar{a}_i\}_{i\in\mathbb{N}}$, $\{\bar{b}_i\}_{i\in\mathbb{N}}$ and a formula $\varphi(\bar{x}, \bar{y})$ of $\mathcal{L}_T$ such that for all i and j in $\mathbb{N}$

$$\mathfrak{M} \models \varphi(\bar{a}_i, \bar{b}_j) \iff i < j.$$

No theory with the order property is $\aleph_0$-stable. This can be seen by applying the compactness theorem to the model $(\mathfrak{M}, a_i, b_i)_{i\in\mathbb{N}}$ as in the definition above, to get an elementary extension $(\mathfrak{N}, \bar{a}_p, \bar{b}_p)_{p\in\mathbb{Q}}$ such that

$$\mathfrak{N} \models \varphi(\bar{a}_p, \bar{b}_q) \iff p < q,$$

and then apply the argument similar to the one we used for DLO.

EXERCISE 15.8. Complete the details of the sketch of the argument above. HINT: See the proof of Theorem 10.3.

We will finish this section with a proof of a result that will allow us to give examples of $\aleph_0$-stable theories.

THEOREM 15.9. *If a theory T is κ-categorical for some uncountable κ, then it is $\aleph_0$-stable.*

PROOF. Suppose that T is κ-categorical and not $\aleph_0$-stable. Then T has a model $\mathfrak{M}$ such that for some countable $A \subseteq M$, $(\mathfrak{M}, a)_{a\in A}$ realizes uncountably many complete 1-types. By the downward Löwenheim-Skolem theorem, we can assume that $|M| = \aleph_1$. By Theorem 15.4, T also has a model $\mathfrak{N}$ such that $|N| = \kappa$ for every countable $B \subseteq N$, $(\mathfrak{N}, b)_{b\in B}$ realizes only $\aleph_0$ types.

By the uppward Löwenheim-Skolem theorem, $\mathfrak{M}$ has an elementary extension $\mathfrak{M}'$ such that $|M'| = \kappa$. Then all the types of $(\mathfrak{M}, a)_{a\in A}$ are also realized in $(\mathfrak{M}', a)_{a\in A}$, but by κ-categoricity of T, $\mathfrak{M}'$ is isomorphic to $\mathfrak{N}$ and this is a contradiction. $\square$

Up to isomorphism, there is only one algebraically closed field of cardinality $2^{\aleph_0}$ and characteristic 0, we get the following corollary.

COROLLARY 15.10. $\mathrm{Th}((\mathbb{C}, +, \cdot))$ is $\aleph_0$-stable.

Also, because every two vector spaces over the field of rational numbers are isomorphic if and only if they have the same dimension, it follows that the theory of vector spaces over the field of rational numbers is $\aleph_0$-stable.

Bibliography

[BS76]   Jon Barwise and John Schlipf, *An introduction to recursively saturated and resplendent models*, J. Symbolic Logic **41** (1976), no. 2, 531–536. MR 403952

[BW18]   Tim Button and Sean Walsh, *Philosophy and model theory*, Oxford University Press, Oxford, 2018, With a historical appendix by Wilfrid Hodges. MR 3821510

[CK90]   C. C. Chang and H. J. Keisler, *Model theory*, third ed., Studies in Logic and the Foundations of Mathematics, vol. 73, North-Holland Publishing Co., Amsterdam, 1990. MR 1059055

[Con18]   Gabriel Conant, *There are no intermediate structures between the group of integers and Presburger arithmetic*, J. Symb. Log. **83** (2018), no. 1, 187–207. MR 3796282

[Doe96]   Kees Doets, *Basic model theory*, Studies in Logic, Language and Information, CSLI Publications, Stanford, CA; FoLLI: European Association for Logic, Language and Information, Amsterdam, 1996. MR 1445772

[Ehr73]   Andrzej Ehrenfeucht, *Discernible elements in models for Peano arithmetic*, J. Symbolic Logic **38** (1973), 291–292. MR 0337583

[EM56]   A. Ehrenfeucht and A. Mostowski, *Models of axiomatic theories admitting automorphisms*, Fund. Math. **43** (1956), 50–68. MR 0084456

[GBGL08]   Timothy Gowers, June Barrow-Green, and Imre Leader (eds.), *The Princeton companion to mathematics*, Princeton University Press, Princeton, NJ, 2008. MR 2467561

[GS66]   Seymour Ginsburg and Edwin H. Spanier, *Semigroups, Presburger formulas, and languages*, Pacific J. Math. **16** (1966), 285–296. MR 0191770

[Hod93]   Wilfrid Hodges, *Model theory*, Encyclopedia of Mathematics and its Applications, vol. 42, Cambridge University Press, Cambridge, 1993. MR 1221741

[Hod97]   ———, *A shorter model theory*, Cambridge University Press, Cambridge, 1997. MR 1462612

[Hod18]   ———, *Tarski's truth definitions*, The Stanford Encyclopedia of Philosophy (Winter 2018 Edition), Edward N. Zalta (ed.), https://plato.stanford.edu/archives/fall2018/entries/tarski-truth/ (2018).

[Kay91]   Richard Kaye, *Models of Peano arithmetic*, Oxford Logic Guides, vol. 15, The Clarendon Press, Oxford University Press, New York, 1991, Oxford Science Publications. MR 1098499

[Kei77]   H. Jerome Keisler, *Fundamentals of model theory*, Handbook of mathematical logic, Stud. Logic Found. Math., vol. 90, North-Holland, Amsterdam, 1977, pp. 47–103. MR 3727403

[Ken18]   Juliette Kennedy, *Kurt Gödel*, The Stanford Encyclopedia of Philosophy (Winter 2018 Edition), Edward N. Zalta (ed.), https://plato.stanford.edu/archives/win2018/entries/goedel/ (2018).

[Kir19]   Jonathan Kirby, *An invitation to model theory*, Cambridge University Press, Cambridge, 2019. MR 3967730

[KM94]   Richard Kaye and Dugald Macpherson (eds.), *Automorphisms of first-order struc-*
          tures, Oxford Science Publications, The Clarendon Press, Oxford University
          Press, New York, 1994. MR 1325468

[Kos11]   Roman Kossak, *What is . . . a resplendent stucture?*, Notices Amer. Math. Soc.
          58 (2011), no. 6, 812–814. MR 2839927

[KS06]   Roman Kossak and James H. Schmerl, *The structure of models of Peano arith-*
          metic, Oxford Logic Guides, vol. 50, The Clarendon Press, Oxford University
          Press, Oxford, 2006, Oxford Science Publications. MR 2250469

[Man99]   María Manzano, *Model theory*, Oxford Logic Guides, vol. 37, The Clarendon
          Press, Oxford University Press, New York, 1999, With a preface by Jesús
          Mosterín, Translated from the 1990 Spanish edition by Ruy J. G. B. de Queiroz,
          Oxford Science Publications. MR 1707268

[Man10]   Yu. I. Manin, *A course in mathematical logic for mathematicians*, second ed.,
          Graduate Texts in Mathematics, vol. 53, Springer, New York, 2010, Chapters
          I–VIII translated from the Russian by Neal Koblitz, With new chapters by Boris
          Zilber and the author. MR 2562767

[Mar02]   David Marker, *Model theory*, Graduate Texts in Mathematics, vol. 217, Springer-
          Verlag, New York, 2002, An introduction. MR 1924282

[MW96]   Angus Macintyre and A. J. Wilkie, *On the decidability of the real exponential*
          field, Kreiseliana, A K Peters, Wellesley, MA, 1996, pp. 441–467. MR 1435773

[Pie17]   David Pierce, *On commensurability and symmetry*, Journal of Humanistic Math-
          ematics **7** (2017), no. 2, 90–148.

[Poi00]   Bruno Poizat, *A course in model theory*, Universitext, Springer-Verlag, New York,
          2000, An introduction to contemporary mathematical logic, Translated from the
          French by Moses Klein and revised by the author. MR 1757487

[Rob63]   Abraham Robinson, *On languages which are based on non-standard arithmetic*,
          Nagoya Math. J. **22** (1963), 83–117. MR 0153569

[Rob66]   ――――, *Non-standard analysis*, North-Holland Publishing Co., Amsterdam,
          1966. MR 0205854

[Rot00]   Philipp Rothmaler, *Introduction to model theory*, Algebra, Logic and Applica-
          tions, vol. 15, Gordon and Breach Science Publishers, Amsterdam, 2000, Pre-
          pared by Frank Reitmaier, Translated and revised from the 1995 German original
          by the author. MR 1800596

[She90]   S. Shelah, *Classification theory and the number of nonisomorphic models*, second
          ed., Studies in Logic and the Foundations of Mathematics, vol. 92, North-Holland
          Publishing Co., Amsterdam, 1990.

[Sim09]   Stephen G. Simpson, *Subsystems of second order arithmetic*, second ed., Perspec-
          tives in Logic, Cambridge University Press, Cambridge; Association for Symbolic
          Logic, Poughkeepsie, NY, 2009. MR 2517689

[Smo91]   Craig Smoryński, *Logical number theory. I*, Universitext, Springer-Verlag, Berlin,
          1991, An introduction. MR 1106853

[TZ12]   Katrin Tent and Martin Ziegler, *A course in model theory*, Lecture Notes in Logic,
          vol. 40, Association for Symbolic Logic, La Jolla, CA; Cambridge University Press,
          Cambridge, 2012. MR 2908005

[Vää11]   Jouko Väänänen, *Models and games*, Cambridge Studies in Advanced Mathemat-
          ics, vol. 132, Cambridge University Press, Cambridge, 2011. MR 2768176

Index

www.ingramcontent.com/pod-product-compliance
Lightning Source LLC
Chambersburg PA
CBHW060353090426
42734CB00011B/2129